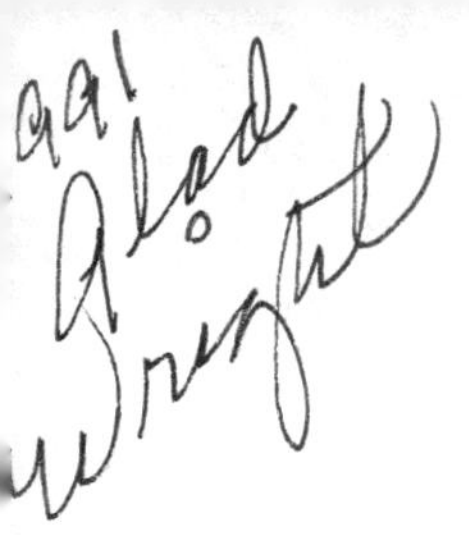

Yeast Disorders

an understanding
and nutritional
therapy

John Finnegan

ACKNOWLEDGEMENTS

This book is the proud result of the collective work of many people. My special thanks to Jeanette Conley, R.N., for her research assistance, editorial input and general moral support; to Daphne Gray for the great time and effort she expended in editing the text; to Lin Winfield for her invaluable editorial insights; and to Robert Natiuk for providing everything from editorial assistance to professional guidance and material support. Thanks also to Phil Van Kirk for his fine graphic work and for managing the production of the book itself. We are all grateful to those who gave their case histories to be used within this work. I especially want to thank my brother Todd who has backed me over the years through thick and thin. And it has been a great pleasure working with such warm, wonderful folks. We all sing praise to the grace which streamed through us in making this creation manifest.

Yeast Disorders
An Understanding and Nutritional Therapy
ISBN 0-927425-00-9

First Printing: April 1987
Second Printing: February, 1989
Published by

20 Sunnyside Ave.
Suite A161
Mill Valley, CA 94941

CONTENTS

LOVE

There is only love
There is nothing to fear

There is no difficulty
That enough love
Will not conquer

No disease that enough love
Will not cure
No door that enough love
Will not open

No gulf that enough love
Will not bridge
No sin that enough love
Will not correct

It makes no difference
How deeply seated
Be the trouble

How hopeless the outlook
How great the mistake

A sufficient realization
Of love
Will dissolve it all

There is nothing to fear
there is only love

John Finnegan
Tony Atkinson

INTRODUCTION

Yeast infections of the digestive tract, sinuses and throat are extremely widespread and constitute the basis of most sinus allergies and many food allergies. More and more frequently, they are being found to be a major cause of hypoglycemic syndromes.

Yeast infections are also a main causative factor in many digestive disorders, cases of fatigue, depression, chronic colds, bronchitis, asthma, premenstrual syndrome (PMS), eczema, and other common conditions.

Dr. Orian Truss's *The Missing Diagnosis* and Dr. William Crook's *The Yeast Connection* have clearly established the reality and etiology of yeast-based illnesses. Even more comprehensive is *The Yeast Syndrome* by Trowbridge and Walker.

The purpose of this book is to introduce a wholesome, simplified dietary therapy, along with some very effective healing modalities, and to synthesize many of the new therapies into a complete recovery program. In addition, I wish to share my own discoveries of a holistic regenerative program that has proven effective time after time.

This approach is not meant to be construed as the only valid perspective, nor as the final word, on the subject of yeast disorders. Each person's lifestyle, environment, background and constitution is unique and healing is a very individual process which can vary greatly.

The basic theme of this book is that there is an incredible natural order in the movement and function of this creation, and the more we understand and live in harmony with it the better we can receive and manifest the love and care that is within this life.

All my best wishes, Finnegan

1

A TWENTIETH CENTURY TALE

The following true case history is typical of the development and manifestation of a disorder that affects far more people than the medical world recognizes today.

"Cathy" enjoyed a very healthy and active life for 50 years. Up until the recent period, described below, a kidney infection was the most serious illness she ever experienced.

I was visiting a friend in Arizona in April 1982, following a four month stay in Costa Rica, when I developed a kidney infection. I had been feeling exhausted since my return a few days earlier, then developed severe back pains and a fever of 105 to 107 degrees. My friend took me to the hospital where I was given antibiotics and sent home.

I asked the doctor if I should curtail my diet in any way, coffee, for instance, which I felt must not be good for kidney infections. He said "No." He even suggested that I drink regular Coke to "get my energy up." Like a fool, I followed his suggestion. Three weeks later I could finally get out of bed and in another two weeks I was able to fly to my home in Berkeley, California.

My diet in Costa Rica had not been "healthy." I stayed with a friend who began and ended each day with coffee and sweet bread and drank beer in between. Most everyone I associated with did the same. I soon adopted the custom. But I also began to eat all the fresh fruit I could to counter balance the effects of the alcohol and sweets—or so I thought. There was an abundance of good, fresh fruit i.e., banana, coconut, mango, papaya, oranges. So I consumed mainly coffee, fruit and sweet breads during the day, then late dinners of beef or fish with rice or potatoes and wine.

When I returned to Berkeley I moved in with friends who ate "good" non-sweet foods. Instead of feeling better, however, I kept feeling worse. It seemed I could not recover from the kidney infection. I felt weak and more tired all the time. I was eating "good" food, at least no more sweets, but still drinking wine and coffee and eating large quantities of fruit.

In November 1983, I went with a friend to his acupuncturist, who thought I might be hypoglycemic and suggested I have a six hour glucose tolerance test. The results indicated low blood sugar. At the same time, I had a complete physical exam. The doctor at the University of California San Francisco Medical Center discounted the results of the glucose tolerance test, saying he did not believe in hypoglycemia. All the other tests showed no abnormalities. So, his conclusion was that I was just depressed and needed an anti-depressant and a good psychiatrist. I declined his kind offer of a prescription and psychiatric referral and continued searching for an answer.

One of my symptoms was extreme fatigue. It was all I could do to get out of bed in the mornings, and some mornings I couldn't even do that. Another was headaches almost every day. Then I suffered from bloating all the time, gas, constipation, a red rash on my face and an increasing sense of hopelessness and despair. It seemed that no matter how much I rested or how well I followed the "hypoglycemic diet," I was always tired, extremely bloated and constantly craving food. So, compulsive eating was added to my other symptoms. I was also gaining weight and felt miserable.

Years ago, I had lived in Bolinas, a small beach community just north of San Francisco. In March of 1984, I decided to return as I sensed that I needed to get back to the ocean and fresh air or I might die.

In Bolinas there were days when I felt better. But for no apparent reason, I would often experience a sudden "decline," and be too exhausted to work or play. I was also becoming aware of more and more moments of confused thinking and strange memory lapses, which I found to be very frightening. During the week, it took all my energy to earn a living. After work and on the weekends I often collapsed into bed as soon as I got home in an attempt to get rested for the coming day or week, too tired to even feed myself.

I was feeling pretty hopeless when I met a nutritional counselor in November 1985. Familiar with Candida and having worked with it for some time, he recognized my symptoms as those of the typical Candida sufferer. He started me on a program to help cleanse my body of this yeast infection. In addition to the diet, he suggested supplemental vitamins and foods.

In a short time I noticed a great improvement, but I still didn't feel like my old self. There seemed to be a veil separating me physically, emotionally and mentally. The intestinal bloating would return, seemingly unrelated to anything I had eaten. I still went through periods of extreme food craving and had fits of compulsive eating, followed by several days of exhaustion.

It was the following August when this nutritionist introduced me to Sunrider products. He said these products were bringing amazing results to people with symptoms similar to mine. With the first sip of Calli tea, I was thrilled! I loved it!

At this time I occasionally drank decaffeinated coffee as this was one of the things I would crave. Whether I drank it or not I had the bloating and other symptoms, so I was too discouraged to even care.

After three days of Calli tea, I no longer even thought of coffee, much less craved it. About that time I experienced some detoxifying symptoms: fatigue, sleepiness, nausea, and a sense of throwing off some kind of chest infection. For four days I also had a constant pain in my lower back and kidney area. But after another four or five days I was feeling fine.

Then I tried the Fortune Delight tea. My body

reacted as if it were a parched piece of earth that at last was being watered. It was an incredible experience! I knew that at last I had found the direction in which I needed to go in order to restore my body's health, which had been completely lost since 1982.

I started taking the Prime Again formula and continued to feel better and better. All symptoms disappeared.

After about a month, I began using a popular protein drink. Here, I had obviously made a mistake. I took it with watered-down fruit juice. The bloating and gas returned. I stopped the juice, then the protein drink. Two days later all my discomfort was gone. I believe I am sensitive to soy protein as I had a similar experience with soy products in the past. It was also a major mistake to use fruit juice as this is a favorite food upon which yeast thrive.

In spite of the bloating and discomfort for those few days, I was feeling so much stronger and better that I was able to handle it, and step-by-step figure out what to do— how to proceed. Hooray! I could think again! A year ago I would have just gone to bed and felt like giving up.

I continue to feel stronger and better, like my "old self." Friends now tell me how good I look and sound. If I am tired from a couple of strenuous days of work, I can rest and be recuperated by the next day. I am no longer constantly hungry and I'm eating less and feeling satisfied with smaller amounts of food. I have not had a compulsive eating orgy for eight weeks and have lost inches as well as weight.

One day I felt stirrings of those old compulsive eating habits. I was very tired from a long week of extra stressful work and other activities. But I was able to remember that I could meditate and rest instead of eat — which I did — and the urge passed.

I now definitely experience a renewed sense of well-being and optimism as well as a return of my ability to think clearly, constructively, even creatively.

I again enjoy reading books about my work and no longer feel the need for escape literature in order to get through exhausting and sleepless nights. I have even signed up for and attend two classes, one in flamenco dancing and the other in mime. I have the energy and enthusiasm for the classes as well as for the daily practice they call for.

In addition, I rejoined the staff of Rites of Passage. I had been forced to quit in 1982 due to lack of energy, stamina and other symptoms. We work with young people (often semi-delinquents and former drug abusers) and adults in transition, to help them in their personal search for meaning and self-direction.

Six months ago, I couldn't have written this self history. I would have been too tired. It would have been too much trouble. I wouldn't have been able to get my thoughts together, and anyway, at the time I thought to myself, who would have cared?

Naturally, I am very thrilled and extremely grateful for the changes in my life. I want to learn more about the many other Sunrider Chinese herbal food concentrates.

Thank you, Sunrider, and also many, many thanks to the nutritional counselor who led me to them and never gave up on me.

2

CAUSES OF YEAST DISORDERS

The body uses four principal mechanisms to control the ever-present yeast populations and keep them from invading and damaging normal body functions.

The first line of defense is to keep a healthy tissue integrity. This consists of maintaining strong cell membranes on the outer skin and inner mucocutaneous tissue linings, such as mouth, sinuses, throat, bronchials, digestive tract, and vagina.

This tissue resistance is highly complex and comprises many interrelating factors. The main factors are hormonal: proper levels of adrenal, gonadal, and thyroid hormones, immunal defenses; and nutritional: optimal levels of proteins, vitamins, minerals and fats.

The second mechanism by which yeast populations are contained is the large, competing population of bacterial floras that are beneficial to the body. By their presence and the substances they produce, these floras keep the pathogenic yeasts, bacteria and parasites under control.

The third main mechanism of control is the acid medium of the tissue fluid which covers the surface of the skin and mucocutaneous membranes, especially in the digestive tract. Pathogenic yeasts, bacteria and parasites can thrive only in an alkaline environment. This was aptly demonstrated in the book, *Folk Medicine* by Dr. J. C. Jarvis. By keeping an acid fluid on the mucous membrane and skin surfaces, the body presents a strong first line of defense to prevent colonization by these invaders.

The fourth bodily defense against yeast overgrowth is only beginning to be understood. This is the basic metabolic and energy cycles within the cells themselves. These metabolic cycles are dependent on many factors: hormone levels, sufficient oxygen, glucose, vitamins, fats, protein and minerals. When these cycles are slow, or lack key factors, resistance of the cells and tissues to pathogenic organisms starts to deteriorate.

A prime example of this function has been repeatedly demonstrated in a laboratory test of healthy cells which were deprived of oxygen and subsequently became cancerous.[1]

The widespread, excessive use of antibiotics is unquestionably the main cause of the current epidemic of yeast infections. Flagyl and other antiparasitic drugs are also a major cause of yeast overgrowth; many of them are actually far more damaging to the tissue lining of the digestive tract and more destructive of the normal flora balance than antibiotics.

Other principal causes of yeast infections include:

Environmental poisons: Indoor pollution. The worst sources are fumes from gas stoves and gas heating, new carpeting and new paint; particle board in furniture, cabinets and paneling; foam mattresses, pillows, and furniture and insecticides. Radiation poisoning from the atmosphere, from groundwater in some states and from medical therapies. Air pollution from automotive vehicles and industrial emissions.

Drugs: Alcohol, caffeine, nicotine, marijuana, LSD, cocaine, amphetamines, etc.

Medications: Birth control pills, auto-immune drugs, cortisone, toxic side effects from other medications.

Dietary Factors: Diets excessive in carbohydrates (especially refined carbohydrates). Antibiotics in meat and poultry, chemicals in foods, sugar, and long-standing protein and trace mineral deficiencies.

Constitutional Factors: Lack of digestive strength, specifically low stomach acid. Liver damage and malfunction. Spleen-pancreas weakness. Under functioning and malfunctioning endocrine and reproductive glands. Lack of floras and colostrum antibody factors given during nursing after birth. Mercury poisoning from silver dental fillings.[2]

Transmission: Passed from mother to child at birth. Sexual relations with a partner who has a genital yeast infection, transmitted from men to women as well as from women to men.

Drugs, nicotine, caffeine, environmental poisons and mercury poisoning from silver dental fillings are not direct causes of yeast overgrowth in the way that antibiotics and sugar are. The first group aid the development of this disorder by poisoning and weakening the glands, immune system, and organs, and by slowing down and interfering with the metabolic cycles.

This is not to say that a healthy person who drinks an occasional coffee, or is exposed to other toxins, will necessarily develop the illness. But if there is excessive exposure, or if one's health is already weakened, these factors can cause the breakdown of bodily resistance. Then the circumstances become conducive for a yeast-based illness to develop.

Psychological oppression and exploitation is also a serious and widespread cause. Clinical tests have demonstrated that these cause physiological damage to both human and animal subjects.[3]

In one test the immune system's T-cell count was found to decrease significantly when people and animals were given work or living situations in which they no longer had control and were subject to external domination.

Physical and emotional illness and repeated financial difficulties are often the result of accepting oppressive life conditions or, on the other hand, from oppressing others.

It is significant that twice as many women as men have yeast-related illnesses. While this is partly due to hormonal and anatomical differences, often a main influence for women developing this condition is that they are of an oppressed class and are not given the respect and love that is their basic right.

Another yeast problem is becoming increasingly prevalent. This affects the gastro-intestinal tract, the throat and sinuses, and often results in rashes in infants. The cause can be traced to mothers lacking the proper intestinal floras to pass on to their children through their milk.

Children who are not provided sufficient floras by their mothers during nursing become easy prey for yeast overgrowth. They soon begin the recurring cycle of colds, sore throats, ear infections and diarrhea—which sends them to the doctor for another round of antibiotics that give the yeast an even firmer foothold.

We are now seeing a whole generation of youngsters and teenagers who have had recurrent undiagnosed yeast infections and repeated courses of antibiotics since birth. By the time they are in their early teens, their endocrine glands and immune systems have become severely compromised, and they have chronic allergies, acne, digestive disorders and fatigue.

One nutritional consultant working in this field maintains that every "universal allergic" she has seen was not nursed from birth and was thus deprived of the vital floras and immune-stimulating factors in the colostrum.

Finally, a major cause of transmission is from mother to child during birth, where there is vaginal yeast overgrowth at the time of birth.

---

1. *The Metabolism of Tumors*, by Otto Heinrich Warburg.
2. *Silver Dental Fillings*, by Sam Ziff.
3. *Mind and Immunity*, by Steven Locke.

3

YEAST: WHAT IS IT?

Yeast is a member of the fungi family. Fifty thousand species of fungi have been described, but it is estimated that there are between 100,000 and 250,000 species. This order includes yeasts, rots, smuts, molds, mushrooms, and mildews.

Their qualifying characteristic is lack of chlorophyll and thus the inability to perform photosynthesis. By nature they are parasitic and saprophytic (feeding off dead organisms) in their manner of obtaining needed nutrients. Fungi, along with bacteria, have an essential function in the recycling phase of ecology. They are responsible for the breakdown of organic matter into its basic elements of oxygen, nitrogen, carbon and phosphorus that would otherwise be bound up forever in the dead bodies of plants and animals.

Parasitic yeast invade living organisms as well, often causing disease or death. Authorities estimate that there are 20 to 30 different types of pathogenic yeasts in the western hemisphere causing disease. Candida alone has 50 to 60 different strains infecting people today. Fungi usually do not live alone. The conditions that foster fungal growth also provide a good home for pathogenic bacteria, viruses and parasites. Yeast invade and infect human tissue in four different ways: superficially, cutaneously, subcutaneously and systemically. The degree of invasion and colonization depends on the amount and strain of the yeast and the strength of the human body's resistance.

Some yeast can, under certain conditions, only affect the superficial areas (skin), while others have the ability to affect the subcutaneous areas. These more invasive strains can ultimately cause systemic infection when the body's condition has been debilitated.

There are probably at least as many infectious diseases caused by yeasts today as there are those caused by bacteria and viruses; yet it is amazing that while the medical profession and general public are so quick to recognize and deal with bacterial and viral illnesses, there are very few physicians who understand and treat the yeast disorders. Meanwhile, yeast disorders are widespread and growing in epidemic proportions. Some authorities estimate that two-thirds of the female population and one-third of the male poplation will have a yeast-based illness at least once in their lifetime.

4

MAJOR TYPES OF YEAST DISORDERS AND ASSOCIATED ILLNESSES

Yeast infections, by themselves and in association with other bacteria, are the cause of most sinus allergies, food allergies, boils and acne, eczema and rashes, vaginal infections, chronic or recurrent kidney and urinary tract infections, and hypoglycemia.

A large percentage of chronic digestive disorders is caused by yeast overgrowth, frequently accompanied by bacteria and parasites. In addition, chronic fatigue, depression and mental disorders are often caused by a yeast infection of the digestive tract.

Premenstrual syndrome (PMS) is very often yeast related. Recurrent colds and bronchitis are frequently caused by a yeast overgrowth of the digestive tract. When the immune system is suppressed, one becomes susceptible to cold viruses and bacteria.

Yeast has a symbiotic relationship with Staphylococcus aureus and other bacteria. Yeast infections in the digestive tract wear down overall immunity, making the body highly susceptible to chronic bacteria — (like S. aureus) infections. Those who suffer recurrent S. aureus infections and children with recurrent ear and throat infections are usually victims of this syndrome.

In several medical studies, animals were first injected with a small amount of yeast-produced fluid, then with live S. aureus. Control animals without the yeast fluid received the second injection and developed a slight inflammation. Their immune systems overcame the staph invasion. But all the animals receiving the slight amount of yeast-produced fluid (Candida) died from severe staph infection. The medical researchers concluded that the yeast-produced fluid is strong enough to destroy the body's immune resistance to staph and make one literally thousands of times more susceptible to staph infections.

The studies also reported that a minimal dose of Candida Albicans greatly enhanced bacterial infections with serrvatia marcescens and Staphylococcus faecalis which can be as harmful to humans as S. aureus.[1]

Widespread outbreaks of yeast infections and associated illnesses are also greatly increased among livestock, due to the constant use of antibiotics in the feed. This is frequently documented in the agricultural magazine, Acres USA.

---

1. Eunice Carlson, "Synergistic Effect of Candida Albicans

and Staphylococcus Aureus on Mouse Mortality", Infection and Immunology, December 2, 1982; 38: 921-924.

Eunice Carlson, "Enhancement by Candida Albicans of Staphylococcus Aureus, Serratia Marcescens, and Streptococcus Faecalis in establishment of Infection in Mice," Infection and Immunology, January 1983; 39: 193-197.

Eunice Carlson, "Effect of Strain of Staphylococcus Aureus on Synergism with Candida Albicans Resulting in Mouse Mortality and Morbidity," Infection and Immunology, October 1983; 42: 285-292.

5

DIAGNOSIS OF YEAST DISORDERS

The most accurate and dependable method of diagnosing yeast disorders is through observation, taking a history of symptoms, then implementing a trial recovery program and observing the results. Because of stereotypical features and the historical development of causes and symptoms, a trained and experienced practitioner can distinguish the yeast syndrome profile accurately in the majority of cases.

A stool culture which measures the level of yeast content is sometimes a good method for determining the evidence of yeast overgrowth of the digestive tract. Occasionally, antibody blood tests or skin tests are useful for confirmation of the condition, but their application is limited. In North America, authorities have documented more than 20 strains of pathogenic yeast infections of the digestive tract. Also, parasites and bacteria (such as chlamydia) are capable of creating this broad disorder.

The antibody tests are specific for Candida only and don't test for the other yeasts, bacteria or parasites that may be involved. More and more frequently, chronic yeast sufferers have bacteria and/or parasites growing alongside the yeast, contributing to their overall distress.

This diagnostic approach may be disconcerting to those who rely solely on clinical tests for evaluation. However, there are many doctors and other practitioners who are able to get good results by using powers of observation and developing discriminating faculties.

My intention is not to blame everything on the yeast syndrome, nor to advocate an off-the-cuff generalization of illnesses as yeast-based. It is well known however, especially among laboratory technicians, that lab tests for different kinds of yeasts and parasites are accurate in a small percentage of cases. Sometimes the tests are useful, but it is essential to realize their limitations and to integrate other methods of diagnosis and analysis.

6

STANDARD THERAPIES: EFFECTIVENESS & LIMITATIONS

The standard medical therapy for treating mucocutaneous yeast infections consists mainly of giving the antifungal Nystatin, with occasional use of Amphotericin B or Nizoral, plus a dietary restriction of carbohydrates. The usual program given for yeast conditions rarely includes an understanding of the proper balance of the various intestinal floras. It usually ignores the necessity for an increase of digestive acidity. Almost always it avoids the necessity of working with lifestyle and ethics or the use of special formulations to nourish and regenerate the glands and organs.

The standard therapy can be dramatically successful, if the doctor is treating a patient of sturdy constitution whose only invading genus is Candida Albicans. This therapy is often ineffective with weak or malfunctioning immune and glandular systems. The added limitation of the Nystatin-based protocol is that it is effective only against the Candida yeast strain. People are often infected with other strains of yeast as well as parasites and bacteria.

Standard therapy also ignores the importance of reducing environmental stresses that suppress the immune system and break down the body's defenses. Some of these environmental factors include paint odors, fumes from gas stoves and heating systems, carpet fumes, formaldehyde from particle board cabinets and paneling, cigarette smoke, and many other chemicals.

The main advantage of the modalities described in this book is that they have broad spectrum yeast-killing properties and help to restore a healthy digestive tract. They also describe some very effective formulations for providing the nutrients that the glands and organs need to rebuild themselves into strong functioning parts of a healthy organism.

7

DIET

Maintaining a good, nutritious diet is essential in reducing and eventually controlling yeast infections. The diet should include plenty of vegetables, both raw and lightly cooked. All vegetables may be eaten, with the exception of winter squash.

Protein may be obtained from fish, fowl, meat and eggs. Small amounts of nuts and seeds may be eaten — one-fourth to one-half cup once or twice daily. For easier digestion, eggs are best soft-boiled or poached.

In addition to the small amount of fat derived from these foods, butter and cold-pressed oils may be used. DO NOT USE processed oils or margarine. Choose a good quality olive, sesame, safflower or other cold-pressed oil instead. Omega Naturals Flax Oil is a valuable source of key fatty acids.

Salads which include raw chopped cabbage, butter lettuce, onions and garlic are very effective in reducing colonies of yeast and parasites.

For those not allergic to yeast, regular consumption of a good chemical-free sauerkraut, sour pickles and vinegar/oil salad dressing is very helpful. Yeasts and parasites require an alkaline medium for their proliferation. These foods, being highly acidic, create an environment unsuitable for their survival.[1]

Another food which helps control these pathogenic organisms is yogurt. If you have no allergy to dairy products, liberal use of a good quality unsweetened yogurt (preferably homemade from one of the cultures mentioned in Chapter 8) is excellent for lowering the yeast population.

It is very important that one eat well and be fully nourished. Many of the frequently prescribed anti-yeast diets are so limiting that people who follow them become fearful about eating and may develop food neuroses and forms of malnutrition.

The diet in this book limits concentrated sweets and starches, while at the same time provides adequate carbohydrates, calories and nourishment through the liberal use of vegetables, proteins and fat.[2]

As far as practical, you should eat organically grown fresh food and avoid anything treated with chemicals or pesticides. This is very important.

Children and teenagers have high caloric and carbohydrate requirements. Be careful not to be overly restrictive in their carbohydrate intake, as this can cause physical and emotional damage. Simply by eliminating concentrated sweets (sugars, desserts, soft drinks, ice cream, etc.), feeding youngsters well and using plenty of flora supplements, Chinese herbal remedies and acid foods, virtually all cases will respond favorably.

The following foods must be eliminated in order to control Candida yeast infections:

1. Sugar, honey, maple syrup.
2. Dried fruits.
3. Fruit juices, except lemon and cranberry.
4. Fruit — in moderate or severe cases. Others should limit consumption to one or two pieces daily of the less sweet varieties, such as apples, strawberries, watermelon, peaches, cherries, nectarines, lemons.
5. Yeast foods (bread, cakes) and fermented foods (miso, amazake) where there is a yeast allergy.
6. Yams, sweet potatoes and winter squash.
7. Mushrooms, in the case of allergy to fungus foods (this is rare).
8. Milk. For those without dairy allergies, yogurt is fine, as is some raw cultured cottage cheese and butter. Butter is an excellent food for most people. It does not promote yeast overgrowth.
9. Most severe and some moderate cases need to eliminate all grains and potatoes until the yeast infection is brought under control. Others should limit grains and potatoes to small amounts — one-half cup once or twice daily. (The best complex carbohydrates are potatoes, rice, millet, buckwheat, corn, quinoa and amaranth).
10. Most people should eliminate all gluten grains (oats, wheat, rye, barley and short grain rice). Bread is especially troublesome and should be avoided.
11. Beans. They are high in carbohydrates and

very difficult to digest.

12. Most people need to avoid soybean products, although a few do well on small amounts of tofu.

13. Cold foods and liquids should not be taken as they weaken the digestion.

It should be emphasized that this dietary outline is meant to be used as a basic guide and not as a rigid dogma. Each person's dietary needs and tolerances differ, so each must find what works best for his or her own body.

Some people do well on small amounts of grains while others will have to avoid them completely for a short period of time.

Generally speaking, the more severe the yeast overgrowth and resulting illness, the more strict one must be with the diet.

My intention is not to add another diet or doctrine to the world, but simply to share some understandings and observations and offer basic principles to work with. Some people require more grains, while others need more raw vegetables, some need more protein and others more fat. The recurrent notion that any one diet or lifestyle is right for everyone is so archaic it is astounding that it is still kicking around in today's world. Each person needs to get in touch with his own body and seek out what works best for him.

---

1. For an interesting elucidation of this phenomenon, see *Folk Medicine* by Dr. J. C. Jarvis.
2. Those who do not tolerate fats and oils should limit consumption accordingly.

8

BENEFICIAL INTESTINAL FLORA

A normal adult has three types of beneficial intestinal floras: streptococcus faecium, acidophilus and bifidus. These bacteria are essential for human life. They help digest food, produce vitamins, and are a key factor in the body's control of pathogenic yeast, bacteria, viruses and parasites. They also provide other important but as yet unknown functions in the immune system.

Each of the three floras has distinct and critical functions, and an imbalance or deficiency in one can severely affect the functioning of the organism as a whole. For example, bifidus breaks down lactose, or milk sugar; a deficiency causes an inability to digest dairy products. In healthy infants, bifidus colonies make up 80 percent of their beneficial flora.[1]

Abundant colonization of the digestive tract with these three floras is one of the main homeostatic mechanisms the body has for controlling the growth and infection of pathogenic organisms. Proper colonization can be damaged by many things. The most common are stress, negative emotions, poor diet (lack of fiber and use of excessively refined and simple carbohydrates), and use of antibiotics, anti-parasite medicines, steroids, and birth control pills.

Proper implantation of all three floras is a foundation block of any holistic yeast recovery program. The quickest and most effective method is to ingest large amounts orally. The quality and potency varies greatly with the different brands of floras available.

One company has pioneered the development of a coated acidophilus formula. One of the difficulties of building up a good intestinal colonization of floras is that the body's strong stomach acids can kill the floras before they reach the small intestine. This company has developed a method of coating the floras so they are protected from the stomach juices as they pass through the digestive tract. There are several companies making good quality acidophilus, bifidus, and streptoccocus faecium formulas.

Those who suffer severe yeast infection of the digestive tract need to take the floras continuously in order to gain control. This can take a few weeks or up to a couple of years, depending on the individual and the severity of infection.

An additional method is rectal implantation of the floras. Some people respond very well to this method. An important part of any vaginal infection treatment is vaginal implantation of acidophilus. Mix one-half to one teaspoon in warm water. Douche with a 20 percent solution of apple cider vinegar, then implant.

Regular consumption of good quality sauerkraut, pickles and yogurt is another important part of the flora replenishing program.

For those without dairy allergies, the best method of implanting large colonies of flora is to culture your own yogurt from one of the three floras and drink a half cup with one or two meals daily.

---

1. A personal anecdote illustrates the uniquely essential function of the individual flora. The six month old son of a friend suffered chronic problems from birth with digesting his mother's milk, with stomach pains, diarrhea, recurrent ear infections and sore throats. We started him on bifidus flora and all symptoms disappeared within the first day. He no longer spat up the milk, the diarrhea stopped, and he became a much happier baby.

9

CHINESE FOOD HERB FORMULAS FOR REGENERATION

Thousands of years ago the sages of China developed herbal formulas with special extraction and concentration methods that are among the most powerful substances known to regenerate and heal weak, deficient body systems.

Today's medical researchers are finding that elements like organic germanium[1] Coenzyme Q_{10}[2], hormonal precursors, and tissue nutrients are some of the remarkable substances in these formulas that give them seemingly magical properties.

However, there are certainly yet-to-be discovered factors in them, and no isolated individual substances have been found that can duplicate the powerful effects of these concentrated nutrients for rebuilding and balancing disturbed body functions.

Over a period of five thousand years Chinese doctors and sages researched, tested, and developed these formulas with unique extraction and concentration methods to perfect whole food herbal formulas which are unsurpassed in their potency and effectiveness.

These ancient sages had a very simple but surprisingly effective method for healing illness. They believed that most illness was caused by weakness of a particular glandular or organ system. They observed that the human body is composed of seven basic functioning systems: the digestive, nervous, glandular, respiratory, muscular, skeletal and circulatory lymphatic systems. When illness manifested, rather than identifying and labeling tens of thousands of separate symptomatic disease entities and treating them as isolated events, they saw the illness as caused by weakness or imbalance of a total body system. They then fed the person the foods and nutrients that would strengthen and regenerate the malfunctioning system and the disorder cleared up automatically.

This particular viewpoint of understanding and healing illness by strengthening the gland or organ system which governs it is gaining more and more acceptance today. This is due in part to the repercussions of many modern therapies: both the public and medical profession are witnessing the damage from side effects that are often a consequence of symptomatic treatment of disease.

Ivan Illich, in *Medical Nemesis*, and many other authoritative researchers, have repeatedly demonstrated the ineffectiveness of symptomatic treatment. More and more physicians today are turning to wholistic methods of healing and some have taken a similar approach to the ancient sages — of treating disease by feeding and strengthening different glands and organ systems. This may be seen in the works of Henry Bieler, M.D. in *Food is Your Best Medicine,* William McGarvey, M.D. in *The Edgar Cayce Remedies* and Elliot Abravanel, M.D. in *Dr. Abravanel's Body Type Diet.*

The Shao Lin priests studied thousands of different plants and classified them into three basic categories:

1) *Poisonous herbs.*

2) *Medicinal herbs* to be used for temporary relief in emergency situations. (These have dangerous side effects and must be used carefully.)

3) *Whole food herbs* which supply nutrients that feed the glands and organs and help them to regenerate.

For instance, carrots and beets are whole food herbs that feed and help heal the liver; cabbage is a whole food herb that feeds and helps heal the digestive tract.

The Shao Lin monks discovered which food herbs had the greatest nutrients for regenerating a particular body system and combined them into formulas containing unsurpassed healing properties. It must be emphasized that these whole food herbal formulas are made from vegetables like carrots and cabbage and so are perfectly safe to use and work by nourishing the body itself.

The following is a compendium of some of the most potent of these ancient Chinese Formulas. These are exceptional because they are grown using organic methods in very fertile soils which provide the plants with an abundance of nutrients and the proper humus conditions for the roots to assimilate them. They are picked at a specific time during growth when the healing properties are at the greatest potency. They are then mixed according to the definitive formulas developed by the Shao Lin Priests through centuries of painstaking research and experimentation.

A primary factor in what makes these formulations so uniquely effective is the specialized extraction process developed for each herb. Many of the nutrients in herbs are bound up in the cellulose, in the complex protein molecules, and in other molecular combinations (fats, minerals) within the herbs. Ordinary boiling and other commonly used extraction methods only make available a small fraction of the herbs' healing elements. It is through the highly complex and specific extraction techniques developed by these priests that the nutrients are separated from their bindings with other substances and made highly assimilable, even by weak and malfunctioning digestive systems.

Special methods were then developed to concentrate the extracted substance to one-eighth its original size thereby making the formula many times more potent and effective.

All the ways in which these Chinese formulas help to regenerate the body will probably never be understood, but four principal functions are known:

1. They give concentrated nutrients that are readily assimilated by the glands and organs for use

in tissue repair and regeneration.

2. They provide precursors for the glands and organs to use in their manufacture of key hormones and enzymes.

3. They provide basic elements (such as organic germanium) which release energy and oxygen at atomic and molecular levels.

4. They provide substances which feed the basic energy cycles, such as the Krebs cycle, in the body's metabolic processes to liberate energy, heat and oxygen.

---

1 K. Asai Ph.D., *Miracle Cure Organic Germanium*
2 E. Bliznakov M.D., *The Miracle Nutrient Coenzyme* Q_{10}

The Chinese Food Herb Formulas

Formula 1

TISSUE CLEANSING & ENERGIZING FORMULA

This is a very powerful deep tissue cleanser. It has been known to dissolve arthritic deposits, help break down plaque in hardened arteries, leach out heavy metals, break up cysts, help neutralize free radicals, and help remove parasites. It also has antifungal properties. It improves liver function, aids fat metabolism, increases energy, strengthens digestion, and improves nervous system function and mental clarity. This is a very pleasant tasting tea. Many people find that it is not only very beneficial to their health but enjoyable as well.

Begin by steeping one tea bag in a quart of boiled water for five minutes. Then start with one half to one cup per day after mealtime. Use only during the day at first. Once you have developed a feel for using this tea, experiment with concentration and amount.

Caution: People with large amounts of stored drug or toxic deposits, or the severely ill should begin using the tea in very dilute amounts. It can cause strong cleansing reactions in some people.

Ingredients: Camelia leaf, Perrilla leaf extract, Mori bark extract, Alisma root extract, Cassia seed extract.

Formula 2

DIGESTIVE CLEANSING & ENERGIZING FORMULA

This formula cleanses the digestive tract, kills yeast, improves digestion, and aids fat and cholesterol metabolism. It nourishes the nervous system and increases energy and stamina. People often experience slight nausea and intestinal upset initially as it kills yeast colonies and cleans toxic matter from the digestive tract. Begin by mixing 1 tsp. of powdered tea in 1 quart of water, and drink one to two cups a day. Increase the concentration of the tea and the amount you use as your system adjusts to it. (Many people use 1 to 3 quarts a day and find that it greatly enhances their sense of well being.) This tea has similar cleansing properties to Formula 1 and thus the same directions and cautions for use apply. If one experiences excessive discomfort, this can be mitigated by: using a more dilute amount, mixing it with dilute Formula 1, or taking it with food. (There is a trace amount of caffeine in Formulas 1 and 2 but not enough to cause harm to the body. [The amount is less than one-half of 1%.] Most of the caffeine is extracted in the processing.) The two teas can be mixed together to produce a beverage with much greater and more complete cleansing and metabolic effects.

Ingredients: Camellia extract, Lemon extract, Chrysanthemum flower extract, Jasmine extract, Lalang grass root extract, King fruit extract and other herbs as flavoring.

Formula 3

NOURISHING, STRENGTHENING, REGENERATING FORMULA

Formula 3 is the most powerful herbal rebuilding formula known. It has a strong regenerating effect on the adrenals, reproductive glands, liver, pancreas, kidneys, and nervous system. It helps build up lean muscle tissue, improves fat metabolism and greatly increases stamina and energy. Athletes use it in large amounts to improve performance, and people convalescing from illness find it substantially aids their recovery.

It is the best formula I have seen for healing the mucous membrane lining of the digestive tract. One woman who was scheduled for a colostomy after fifteen years of deteriorating ulcerative colitis began using this formula, along with the other formulas, cancelled her operation, and recovered in a few weeks.

People with severe fatigue or blood sugar problems, and those recovering from addictions will find that it greatly stabilizes their metabolism and increases their energy when they use it several times a day. It is an excellent food for infants, as well, and will build up their health.

Formula 3 mixes up best in a blender. Maintenance usage is two tablespoons once or twice a day. For building strength or recovering from an illness, use two to three tablespoons three times daily. Mix it in the Tissue Cleansing & Energizing Formula 1, and take it with Metabolism Regulation Formula 10,

Endocrine Gland Formula 6, Korean Ginseng, and Women's Glandular Formula 7.

For weight management, mix it in Digestive Cleansing & Energizing Formula 2 with Metabolism Regulation Formula 10, and take a glass one-half hour before each meal.

It also makes great smoothies. Take this formula one to three times a day, and you will experience significant results. This formula is made entirely from organically grown herbs that are extracted and concentrated five to eight times.

Ingredients: Coix Fruit, Chinese Yam, Fox Nut, Lotus Seed, Lotus Root, Apple, Water Lily Bulb, and Imperate Root.

Formula 4

HEALTH STRENGTHENING FORMULA #1

Helps rebuild and improve immune system function. It will strengthen the body's natural defenses and enable it to better overcome colds, viruses, and other forms of infectious illness.

The immune system and the glandular system are the body's main mechanisms for preventing and recovering from illness. Certain foods and food herbs provide nutrients which greatly improve immune system function. This formula, when taken with the Health Strengthening Formula #2 and Endocrine Glandular Formula have shown remarkable success in regenerating and improving the body's overall health. Two to six capsules daily helps prevent or heal mild illnesses. Up to 16 capsules daily is recommended for severe or chronic conditions. Very severe cases can use 25 to 50 capsules a day for a short duration to aid recovery.

Ingredients: Herba Menthae, Flos Lonicerae, Rhizoma Ligustici Wallichii, Fructus Forsythiae, Herba Schizonepetae, Radix Platycodi, Rhizoma Notopterygii, Radix Angelicae, Radix Glycyrrhizae, Folium Phyullostachys, Fructus Arctii, Rhizoma Phragmitis.

Formula 5

HEALTH STRENGTHENING FORMULA #2

Feeds the immune system and aids increased production of T-cells. T-cells are immune system cells that fight infection and play a major role in regulation of allergic response to invading organisms. Using this formula to nourish and strengthen the T-cell production often results in an improved capacity to fight infections, identify and break down growths and normalize allergic reactions. Many people with severe immune system disorders have experienced very effective help in recovering by using the two health strengthening formulas in large amounts: the specific ailments include chronic mucocutaneous Candida Albicans infections of the digestive tract, throat, sinuses, vaginal areas and skin, and others with universal reactor syndrome, chronic respiratory disorders and immune dysfunction conditions. Recommended dosage is the same as for Formula 4.

Ingredients: Chinese white flower, Paris herb, Scutellania herb, Dandelion, and Bai-mao root.

Formula 6

ENDOCRINE GLAND FORMULA

This preparation feeds and improves functions of the endocrine glands, including the adrenals, thyroid, and reproductive glands. These glands produce hormones which are essential to life and regulate all the body's major functions and energy cycles. Among these are the blood sugar level, protein assimilation and muscle building, nervous system function, mental well-being, immune system function, detoxification of poisons, digestive processes, body stamina, body temperature and other essential body processes.

Many illnesses, from arthritis to a susceptibility to recurrent infectious diseases and many vague bodily disorders such as fatigue, overweight, poor digestion and poor nervous system strength, are caused by weak or underfunctioning glands. Strengthening and improving their function often clears up a whole host of disorders and dramatically improves overall body health and personal enjoyment of living. This has been well documented by many authoritative researchers including: Broda Barnes, M.D. in *Hypothyroidism: The Unsuspected Illness*, Stephen Langer, M.D., in *Solved: The Riddle of Illness*, J.W. Tintera, M.D. in *Hypoadrenocorticism,* Ray Peat, Ph.D. in *Nutrition for Women*, and others.

This Endocrine Formula, as well as the Women's Glandular Formula, the Korean Ginseng Extract Formula, the Tissue Cleansing and Energizing Formula 1, the Metabolism Regulation Formula 10, and the Nourishing Formula 3 have all demonstrated safe, repeated success in feeding and improving function of the endocrine glands. Irregular menstruation, PMS and menopausal symptoms have been greatly helped when taken along with the Women's Glandular Formula. Recommended dosage is three to six capsules a day for mild to moderate conditions and up to 12 daily for severe conditions. Pregnant women should not take more than one capsule a day, as it might contribute to the baby's development as unusually large and cause difficult delivery.

Ingredients: Radix Dioscoreae, Radix Cyathulae, Herba Epimedium, Semem Allium, Poria, Fructus Corni, Fructus Broussonetiae, Cortex Eucommiae,. Fructus Schizandrae, Fadix Morindae, Herba Cistanches, Radix Polygalae, Fructus Foeniculi, Rhizoma Acori Graminei.

Formula 7

WOMEN'S GLANDULAR FORMULA

This formula strengthens the endocrine glandular system and nourishes and improves the growth and healing processes. It specifically feeds the female hormonal system and often has a remarkable effect on improving PMS and menopausal symptoms. It increases energy and helps calm and stabilize the nervous system.

A pleasant side effect of strengthening the body's glandular systems is the improvement of skin tone, complexion, and beauty. People often say that their faces become ten to fifteen years younger looking within a few months of using this formula, along with the other formulas.

One mother's experience that demonstrates the capacity of this formula to nourish overall bodily regeneration happened when her sixteen month old child had been ill and failed to grow at all in the previous 6 months. The mother cut the tablet into little slivers, put them on his almond butter sandwiches, and in the next month he grew an inch.

Many people find this formula to be very strengthening and healing. For maintenance purposes use one tablet daily with meals, and for more therapeutic uses take 2 tablets daily.

Formula 7 contains extract of real pearl, which has traditionally been revered in the Orient as a source of special elements that rejuvenate the body's vital processes.

Ingredients: Honey, Korean Ginseng, Chrysanthemum Flower Extract, Royal Jelly Extract, Pearl, and other herbs as flavoring.

Formula 8

DIGESTIVE SYSTEM FORMULA

Feeds and helps heal the digestive tract. Cases of poor assimilation, lack of enzymes, excess or insufficient hydrochloric acid, and intestinal irritability can be greatly improved with the use of this formula. This formula doesn't work by providing additional digestive enzymes, though many digestive remedies take that approach. Instead, this formula feeds and strengthens the stomach and the supporting digestive organs, so that their production of enzymes and acid secretions can be normalized and restored. Three to six capsules daily for mild to moderate conditions, and up to 12 per day for severe or chronic conditions.

Ingredients: Radix Ginseng, Rhizoma Atractylodis, Poria, Radix Glycyrrhizae, Rhizoma Pinelliae, Citri Reticulatae, Cortex Cinnamomi, Rhizoma Corydalidis, Fructus Foeniculi, Fructus Amomum, Herba Methae.

Formula 9

CIRCULATORY SYSTEM FORMULA

Poor circulation is a factor in many illnesses. A decreased blood supply to the brain and nervous system will cause depression and poor memory. A diminished blood supply to body tissues and organs will cause a decreased amount of oxygen and nutrients, and a buildup of toxins which will create the conditions favorable for all types of disease to develop.

This formula increases vascular elasticity and helps heal circulatory disorders including varicose veins, hardening of the arteries, blood pressure malfunctions, excessive fat and cholesterol levels, and will also improve eyesight. Use three to six capsules daily for mild to moderate conditions and prevention of illness. Use up to 16 capsules per day for severe or chronic conditions.

Ingredients: Semen Cassia Tora, Uncariae Ramulus, Flower Sophora, Citri Reticulatae, Rhizoma Pinelliae, Radix Ophiopogonis, Poria, Rhizoma Zingiberris, Radix Ginseng.

Formula 10

METABOLISM REGULATION FORMULA

A very powerful preparation with broad effects on many body functions. Excellent for weight management, it helps restore normal fat metabolism function. Many people report not only a loss of excess fat but a building up of lean muscle tissue as well. It picks up stamina and overall body energy, is a superlative cleanser of the digestive tract, and also improves the spleen-pancreas, liver, digestive, and glandular functions. This formula is also great for building up muscle tissue and strength in those who are underweight. Formula 10 consists of three separate formulations, to be taken together two to three times daily in a dosage of two to three capsules of each according to individual need.

Ingredients: Chinese yam, Taro powder, Plantago Asiatic, Camellia leaf extract, Bai Mao root extract, Caulis Hocquartial extract, Rhizoma Alismatis extract, Brigham tea extract, Senna seed extract, Rehmannia Lutinosa extract, Cortex Mori Radicus extract and other herbs as flavoring.

Formula 11

HEALING BALM

Developed by the Shao Lin monks to aid recovery from injuries sustained during martial arts training. It was later found to be helpful for numerous other conditions, including insect bites, back pain, stiff neck, sore throat, heavy chest due to cold or flu, toothache, and sinus congestion. One nurse had a patient with a severe strep throat that was not responding to antibiotic treatment. She applied the balm to the sides of his neck, and within minutes there was tremendous improvement. For headaches apply to temples and back of neck. For sprains, bruises, minor arthritis and bursitis pains, rub into affected areas. Formula 11 also helps heal cuts, burns, sunburn, and minor infections. Many people get relief from motion sickness by putting a bit on the tip of the tongue and the back of the neck.

Ingredients: Menthol, camphor (2-5%), Cassia oil, plus other oils as fragrance.

Formulas 12, 13, & 14

NERVOUS SYSTEM FORMULAS

Scientific research today is turning up evidence supporting what the Chinese sage doctors knew thousands of years ago. Recent studies on the health and functioning of the nervous system have discovered that certain key nutrients, neurotransmitters, prostaglandins, and hormones are essential to normal nervous system perception and function.

Stress, poor nutrition, poisons and genetic impairment can all contribute to creating a deficiency of these substances which will, in turn, cause serious malfunction of the nervous system.

The works of Dr. Fox in *DLPA*, Judy Graham in *Evening Primrose Oil*, Dr. Langer in *Solved: The Riddle of Illness,* and others have clearly shown the critical effect that these substances can have in clearing up depression, fatigue and phobias, and in alleviating the pain of arthritis and other diseases by strengthening nervous system function.

Today's pharmaceutical pain killers work by deadening the nervous system, and all have certain side effects. Proper use of nutrients and food herb formulas work by nourishing, normalizing and increasing the capacity of the nervous system to function well under stress.

While there is clearly a place for using individual substances in improving nervous system function, the most important thing is to provide complete nourishment of all essential nutrients through the use of whole foods and whole food extracts.

The importance of calcium in maintaining normal nervous system functioning is well known, but what is not so well known is the difference in absorption and utilization of commonly used supplemental calcium products as opposed to calcium taken from whole foods.

A recent study on human absorption showed that test subjects failed to absorb 93% to 100% of calcium from calcium carbonate and oyster shell.

This same principle of the significant difference in absorption and utilization of nutrients from foods and food concentrates, as opposed to isolated or synthesized nutrients, applies to most nutrients and is a major reason why foods and food extracts have such superior healing properties.

Since the endocrine glandular system (mainly the thyroid, adrenals, and reproductive glands) contributes at least 50% of the strength of the nervous system, the basic program should provide ample amounts of the Nourishing, Endocrine, and Women's Glandular Formulas along with the Cleansing teas #1 and #2.

GENERAL NERVOUS SYSTEM FORMULA #12

This is the basic formula for strengthening nervous system function. Its use is indicated in conditions of general pain, burn out, and arthritis. It is used to aid mental concentration and clarity. Dosage is 2 to 3 capsules once or twice daily, and more often during serious illness.

Ingredients: Cassia Tora Seed, Gou Teng, Ji Tsau Herb, Sophora Flower, Yeuan Wu Root, Orange Peel, Pinelliae Root.

MENTAL CLARITY FORMULA #13

This formula is used with Formula 12 to enhance mental acuity. People who work with computers and bookkeeping have reported that by using 2 or 3 capsules of #13, along with 2 capsules of #12, they can work twice as long without mental fatigue. This formula is also good for headaches, neck and shoulder tension, insomnia, and mental difficulties.

People with allergies have reported that using Formula 13, Formula 12, both Immune System Formulas, and the Endocrine Formula has cleared up their allergic conditions. Besides giving immediate relief, continued use over a period of time will help the body to heal the underlying disorder and aid permanent recovery. Dosage is 2 to 3 capsules once or twice daily, and more often during serious illness.

Ingredients: Mint, Silver Flower, Chuan-Xiong Root, Yeuan Wu Root, Angelicae Root, Golden Bell Fruit, Ji Tsau Herb, White Willow Bark.

JOINT & INJURY FORMULA #14

This formula is used for pain from injury, trauma, and joint deterioration. It is usually used with Formula 12 to give supportive nutrients. Research into the function of neurotransmitters has shown how critical it is to have adequate nervous system nutrients to heal injury and clear up pain. Dosage is 2 to 3 capsules once or twice daily and more often during serious illness.

Ingredients: Siberian Ginseng, White Willow Bark, Mint, Silver Flower, Yeuan Wu Root, Chuan Xiong root, Angelicae root, Golden Bell Fruit.

Formula 15

THROAT LOZENGES

The mouth, throat and sinuses can harbor the growth of many types of pathogenic bacterias, viruses, and yeasts. These can cause everything from sore throats and thrush, to tooth decay and sinusitis. The ancient Chinese herbalists knew what modern medical researchers are just discovering. Different plant oils have very powerful anti-bacterial, antiviral and antifungal capacities. These oils also strengthen the tissues themselves. Singers and public speakers say they can sing and speak twice as long when using these lozenges. Use as needed.

Ingredients: Gum Acacia, Sorbitol, Mint, Thyme, Eucalyptus, and other herbs for flavoring.

ADDITIONAL SUPPORTIVE HERBS

GOLDENSEAL

This herb is superior for cleansing the liver and bloodstream. Goldenseal was considered to be the finest blood purifier and herbal antibiotic by such renowned authorities as Jethro Kloss, Dr. John Christopher, and several others.

Its use is especially indicated when one is healing excessive gut permeability caused by candidiasis. This breakdown of the intestinal mucosa is often a major causative factor of food and environmental allergies.

Goldenseal is an excellent herb to take during an infectious illness. It has been used effectively against respiratory infections, genitourinary infections, and against gastrointestinal diarrheas caused by E. Coli, Shigella Dysenteriae, Salmonella Paratyphi B., Giardia Lamblia, and Candida Albicans. Both human and culture studies have shown it to have significant amoebicidal activity against Entamoeba Histolytica. It has also been found to be active against Staph Sp., Strep Sp., Chlamydia Sp., and Trichomonas Vaginalis. Goldenseal's action against some of these pathogens is actually stronger than that of the antibiotics commonly used.

However, unlike prescription antibiotics, Goldenseal does not kill the beneficial digestive floras while it does kill candida and other pathogenic yeasts. This information was extracted from a summary of 29 medical studies describing Goldenseal's activity in the February 1988 issue of "Phyto-Pharmacia" included in the "Townsend Letter" June, July 1988.

People with infectious diseases, yeast disorders and ulcerated gastrointestinal tracts have found three to six capsules a day for one to three months to be of great aid in healing this condition. More may be taken during acute illness. Goldenseal should not be taken for more than three months nor used by pregnant women.

KOREAN WHITE GINSENG

Ginseng is considered to be the king of tonic herbs. It feeds and strengthens the endocrine glands and overall metabolism. Athletes use Ginseng to increase their stamina, and people convalescing from illness use Ginseng to build up their health. It is used to strengthen the heart and normalize blood pressure. It also nourishes the blood, having a beneficial effect on anemia.

Its use is especially indicated for those with weak adrenal or reproductive glandular function, and those with hypoglycemia, fatigue, and nervous system weakness.

This herb has special catalyzing properties and will increase the effectiveness when used with other strengthening herbal formulas.

Three to six capsules a day is the recommended dosage.

SIBERIAN GINSENG ROOT BARK

This herb nourishes the spleen-liver function and helps strengthen joint tissues, ligaments, and tendons. Traditionally, it has been used for lack of appetite, insomnia, forgetfulness, nervous disorders, low energy, and convalescent weakness. Three to six capsules daily is the recommended dosage.

CHRYSANTHEMUM

This herb is used widely in Chinese herbal formulas. It has powerful antifungal, antibacterial and anti-inflammatory properties. It helps liver function, and traditionally the Chinese used it to correct blood pressure irregularities.

STEVIA

Stevia has a long history of safe and therapeutic use both as an herbal sweetener and as an antifungal, anti-inflammatory and antibiotic agent. It has been used for centuries by the natives in South America and for the last few decades in Japan and Europe, where it is much acclaimed by their medical professionals as a dentifrice and a blood sugar stabilizer.

It is thirty times sweeter than sugar, yet has practically no sugar in it. It has been found to lower the blood sugar in diabetics and raise the blood sugar in hypoglycemics. It is also remarkably efficacious when used topically for poison oak, athletes foot, rashes and infections.

DONG QUAI

This is good for feeding the body's endocrine glands and improving the blood quality. This herb has long been treasured among both eastern and western herbalists. In the east, it is considered one of the best food herbs, and people cook it regularly in soups to increase their health and strength. It is favored by women to strengthen their glandular function, but should not be used during pregnancy. Recommended dosage is three to six capsules daily.

WHITE WILLOW BARK

This herb helps to reduce pain and inflammation and improves the body's resistance to illness. Three to six capsules is the recommended dosage during acute illness.

FIBER

Our diets today contain less than one-third the amount of fiber contained in the diets of our grandparents, and in the diets of people in more rural areas of the world. The average western diet contains 10 to 20 grams of dietary fiber, while the diets of most rural societies contain 40 to 60 grams a day. According to the now classic paper that Dr. Denis Burkitt published in The Lancet in 1969, entitled "Related Diseases-Related Cause?," intestinal diseases which are prevalent throughout the civilized nations are almost unknown in rural Africa and many other societies.

According to Dr. Burkitt, most leading degenerative diseases are partially caused by an insufficient intake of dietary fiber. These diseases include coronary heart disease, diverticular disease, appendicitis, hermorrrhoids, varicose veins, obesity, and diabetes. The second most common cause of death from cancer is cancer of the colon, which has been firmly linked to a low fiber diet. Even Candida overgrowth is associated with low fiber diets. While this information was reported in 1969, and nutritional healers have been saying these things since time immemorial, today we are seeing many articles about using oatbran to reduce cholesterol.

There are five known functions that fiber in the diet performs:

1) reduces intestinal toxicity and pathogenic bacterial and yeastovergrowth.

2) improves bowel functioning and transit time

3) stabilizes blood sugar. It has been proven by many medical tests to help regulate both diabetes and hypoglycemia.

4) lowers cholesterol.

5) protects against other chronic degenerative diseases, such as cancer of the colon, hemorrhoids, varicose veins, etc.. Proper use of fiber has been firmly established to be the most effective therapy for clearing up hemorrhoids.

There are several ways to improve your dietary intake of good quality fiber: reduce consumption of overly refined foods; eat a salad every day with plenty of lettuce, grated carrot and other fibrous vegetables; use a liberal amount of whole grains and legumes in your diet; and, when needed, take some kind of high quality fiber supplement.

CHILDREN & CHINESE FORMULAS:

Development and usage of these food herb formulas for 5000 years have given us products proven to be without toxic side effects. All of the herbal formulas are safe and beneficial for children to use. Many babies and young children have enjoyed improved health and well being through the use of these Chinese herbal preparations. Infants and children especially thrive on the Nourishing Formula.

The Health Strengthening Formulas are especially effective for children. The dosage should be decreased by half for children aged two to ten; and one to two capsules per day is sufficient for infants.

Use of the two cleansing and energizing formulas must be monitored carefully, as it would be hazardous to cleanse too rapidly or increase the energy of children too dramatically. Begin with small, diluted amounts for youngsters and increase gradually. The Nourishing Formula can be given full strength with good result.

10

ON SUPPLEMENTS

In addition to proper diet, there are many vitamins, minerals, enzymes, homeopathic preparations and other supplements that are important in any yeast recovery program. The amounts and kinds needed by each individual vary greatly.

It should be understood that the whole nutrition and vitamin market has become very commercialized. Many products are grossly misrepresented, over-priced, and of poor quality. Some are even harmful.

Many people with digestive tract yeast infections have had the sensitive tissues lining their gastrointestinal tract invaded and eaten away by the yeast colonies. They experience great pain and difficulty absorbing high doses of synthetic B vitamins and may encounter not only severe pain, but also tissue damage from using them. It is essential for these people to begin using vitamins and other formulas with a very low dosage and increase gradually, in order to avoid painful reactions.

Any basic supplement program should include good quality chelated calcium and magnesium. In addition, Vitamins A, D, E and co-enzyme B6, chelated Vitamin C and garlic extract are among the supplements most often found beneficial for healing this condition. The best vitamins are extracted from whole foods. *The Yeast Syndrome* contains a good description of the kinds of supplements beneficial in a yeast recovery program.

Because individual needs vary greatly, it is important to work with a knowledgeable holistic physician when putting together a yeast recovery program.

11

THREE CASE HISTORIES

A 13 YEAR ODYSSEY

"Diane" is in her early twenties. She loves music, which she began to study at age 8. She has lived in Europe, New York and the San Francisco Bay Area. In spite of having spent half of her life dealing with health difficulties, she remains enthusiastic and high spirited.

My health problems, I believe, trace back to infancy. When I was two months old, my mother stopped nursing me because her milk ran dry. As a result, I received little of the friendly bacterial floras usually acquired from mother's milk and which aid in building the foundation of a healthy, robust immune system.

Next in the line of mischief-makers was parasites. Pinworms — who came to live in me when I was 7. Two weeks of heavy-metal medication was administered and I was to have no further problem with them until age 9, and then again at 10.

A perfectly innocent, secluded nuisance, right? Well, so we thought. Indeed, several years of relative peace followed before a whole host of illnesses were to manifest and which — we were later to discover — were all Candida based.

The symptoms that developed were: ear and throat infections, digestive problems, severe female disorders, food allergies and overwhelming exhaustion. It was a struggle just to be alive. To my astonishment, various "psychological" symptoms disappeared in the very beginning of my anti-Candida regime: paralyzing shyness, anxiety attacks, concentration problems, memory loss and psychosis. I had no idea that these were physical symptoms, I thought they were emotional.

The chain of events began with recurrent ear and throat infections, beginning when I was 8 and disappearing finally around age 11, thanks to the help of my homeopathic doctor.

After a year of peace, I developed severe indigestion. There were times when I stopped digesting altogether and could not eat anything at all. Whether these phases lasted for days or weeks, I do not remember. But I do recall that they existed over extended periods of time. I tried several doctors with differing medical approaches.

The first doctor was a regular M.D., whom I had heard was very good. He put me on Maalox and prescribed tranquilizers, suggesting privately to my mother that the problem was psychological. She refused to put me on such heavy drugs, but I continued to see this doctor for a while. He diagnosed me as having a pre-ulcerous stomach condition and hyper-acidity, and that was the end of that.

Later we tried a homeopathic doctor who also prescribed psychological treatment in addition to homeopathic therapy. My mother tells me that she was unable to talk to anyone — doctors, family, or friends — without getting the line that stomach problems were psychological and that I was a sensitive subject, a likely candidate for such troubles. She felt cut off and unable to seek help and did not know what to do for me.

Years later, I developed female disorders, manifesting as polycystic ovaries. My symptoms began one time when I was menstruating. I was sunbathing by the swimming pool at the college I was attending and feeling just great. Suddenly I felt my body go numb and start to tingle. Throbbing, metallic stabs shot around the pelvic area, my whole body racked with pain. I felt my face turn white as a sheet, especially my lips. I nearly passed out, the pain was so intense. The world grew black around me; I couldn't see anything and dizziness seized me. I could hear people talking to me, so I knew I had not lost consciousness.

Such episodes became regular experiences during my monthly menses, with the pain increasing in intensity each time. Over the course of two years, these two minute episodes stretched out to last for days, two weeks, two months, and then finally, three months at a time. It was agony. I couldn't move, eat, cry or go to the bathroom. If I wanted to roll over in bed, it was an excruciating maneuver. Often, just shifting my body weight made me feel as if my insides were going to explode. It was as if someone had planted a knife in my gut and left it there. So tender was my body that if someone placed a hot water bottle on my abdomen a little too quickly or with a bit too much water in it, my insides would throb wildly, sending waves of pain throughout my whole torso.

To reach for the phone and call the doctor at these times was an impossibility. I was forced to wait for such crises to pass before I could go in for a check-up. Neither allopathic nor holistic doctors knew what to do for me, nor could they determine what was wrong.

No one seemed to understand that I really was in pain. According to my present nutritionist, it is quite common for the Candida sufferer to be accused of hypochondria or written off as a psychological "case." This is what happened to me. I remember one emergency room doctor very righteously saying, "Ovarian cysts don't hurt that badly, I've had them." "But this is just a small speculum," said another as I moaned while she inserted it. Similarly my gynecologist, who had a very fine reputation, did not believe or comprehend that I was really suffering. By the time I would be sufficiently through a crisis to see him, the ovarian cysts would have leaked out, leaving little or no trace of their presence. "Polyeptic ovary disease," he labeled my condition, prescribing a 30 year course of birth control pills. In retrospect, I am very glad I trusted my gut instinct not to follow his suggestion. The pill may have masked my symptoms, but made the cause, Candida, even worse.

I began seeking alternative approaches for my female troubles. Having worked successfully with homeopathy in the past, I chose this route to help me. Two years of treatment brought some relief from pain, but as yet no cure.

Over a two year period, I worked with three different homeopathic doctors, all puzzled by my case. "Endometriosis," speculated one about my ovarian symptoms. "Anxiety," said another. "She is physically weak and her nerves are therefore delicate," suggested another. Doctor number two, with whom I worked for a year, tried several remedies but none of them worked. He brushed aside my stories of pain and expressed far more concern about my constantly exhausted state. I would sleep for four hours, be awake for two or less, and then be down again for four more. This cycle went on all day long. My limbs felt like rubber and I felt half alive, half dead. I recall days when the act of walking — just putting one foot in front of the other — was a great effort. But these new symptoms brought blessed relief from the pain which stopped my whole life. I was grateful to be mobile and to once more do my music, which I had been forced to abandon. However much effort it took me to do this was of little consequence. I was mobile, I could do it! This was a joy to me.

Meanwhile, roommates, friends and siblings sat on me for not working like the capable, intelligent person that I was. And who can blame them? When out of bed, I looked absolutely normal and had all my usual exuberance and liveliness. Only my mother, who nursed me through my crises, could have understood that I was seriously ill.

At one point, I began to lose my memory. Then as memory lapses became the norm, I thought I was going senile. At work I found myself unable to do simple calculations in arithmetic, a subject I formerly excelled in. Then I developed nervous disorders. I had irrational, unremitting fears and hallucinations, accompanied by severe insomnia. I felt angry at absolutely everyone for no reason, and believed that a suppressed emotion was causing all my symptoms. Was this early insanity?

I felt certain I was headed for the looney bin. I knew that I desperately needed help. But my angel was watching over me and led me to just the right person. I had virtually everyone praying for me, and after three days of prayer my mother met a nutritional consultant who had almost cured himself of environmental allergies within a matter of months. He used Chinese herbs put out by a company in the United States called Sunrider. What these herbs had done in months, he said, took years for many doctors to achieve for him, including such herbal wizards as Dr. Christopher.

Thank God for this nutritional consultant. He introduced me to the disorder of Candida and described in detail its many symptoms. I felt a stone lifting from my heart as he did so. It was an incredible relief to find that all my symptoms were typical of a disease. Even my supposed "insanity" was due to a physical problem! I had thought my various symptoms represented separate illnesses, requiring many cures, a depressing prospect. Now I found they were all connected to a common cause.

It was even more encouraging to learn that my nutritionist and several of his clients, with worse cases than my own, had recovered. We traced my history back to the metal medication I had taken for pinworms.

The nutritionist and his various clients were having dramatic results with the Chinese herbs. Had I not grown up using herbs, I would never have believed their testimonies. How could a hospitalized Candida patient, expected to die in a matter of weeks, turn around in the same amount of time? I did have trouble believing the story, but I trusted this man as a person and sensed that I could get some benefit from his herbs.

I experienced results literally with the first cup of Calli tea. So did my mother. Within a few days I began digesting my food rapidly and felt constantly hungry — something I had not felt for eight years. My whole abdomen felt much lighter and more clear. I began to lose inches from my belly, since I lost much retained gas. All this happened within a few days.

The next week my sleeping patterns improved. By the third week, I suddenly realized that all my emotional symptoms had completely vanished. Equally encouraging, I noticed that I could now read with greater ease. I no longer had to stop and reread every sentence. I could read a book and enjoy it, and could also think with greater clarity and efficiency.

I felt my old intelligence returning, and all my old interests with it, which had been abandoned for many years. I felt so much more like my real self! Within the first week my exhaustion had improved to a point where I could get through a day without sleeping. Within three weeks, I had all my normal exuberance and liveliness. Only those who know me realize what a statement that is!

After three weeks, I relapsed. My ovarian pain returned. I mentioned this to the nutritionist and added that I would rather have committed suicide than ever go through such an experience again. He assured me that this was not very likely to happen and that, due to the detoxifying properties of the herbs, it was quite usual for old symptoms to temporarily resurface.

He suggested using hot castor oil packs, which were tremendously helpful. Equally so were the acidophilus supplements he gave me to combat the Candida. Of the two, I would say the acidophilus was the greater pain killer, in addition to being something that would help build me up. Within a single day, all pain was minimized to the point where it was mild and easy to bear.

Such crises never came up for me again, although I have had milder periods of detoxifying that brought back some nausea and tiredness. These, however, were to last only for a few days at a time.

After six weeks on Sunrider products I felt like a new person — with energy like never before! The color returned to my face and I lost 15 pounds, in spite of eating like a pig. My elimination eased and became more regular; my periods become practically painless. My hair and skin changed texture, becoming smooth and fine. Digestive problems no longer plagued me and the ovarian pain diminished. In short: I began to feel terrific!

Now both friends and acquaintances tell me how good I look. I feel more productive, more spontaneous, and old shyness is falling away. As I gain more energy, I have better concentration. All in all, I feel I am living a much richer life. I feel a new epoch has begun in my life, and things are working for me in my social life, my love life, and my career. Some of this is the result of the work I have been doing on myself as a person. But most of it — about 70 percent — is the result of my body getting healthier and stronger. I pass this on for the benefit of those who are suffering with some illness in the hope that my experience may inspire hope in others.

I have had several opportunities to see the external effects of Sunrider as well as internal ones. My mother had a bad case of poison oak on the upper and lower lids of her left eye. She applied the Suncare (stevia extract), sprinkling the herbal clay over it. The itching stopped the moment she applied the powder. Within three days, the poison oak was gone and the swelling with it. She said it worked faster for her than Cortisone.

Similarly, we were able to heal my roommate's dog of flea dermatitis by mixing the stevia extract with oil. This was applied to the skin twice daily and the bleeding sores cleared up in a few days. It also cleared the disgusting flaking callouses on his back.

Within two weeks, this dog's hair was growing back. He now looks like he has never been ill in his life. I might add that, some years ago, my own dog had to be put to sleep for this condition. Neither veterinary prescriptions nor professional medical baths had helped her.

I hasten to add that these are not miracle herbs, nor do I wish to imply that my cure has been complete. In addition to faithfully using the Sunrider program, I must also be meticulously careful with my diet, or I regress.

Although the NuPlus and Suncare have greatly reduced my food cravings, temptations still strike me to eat the wrong foods. Twice I gave in to this temptation and twice I relapsed. I have since concluded that Candida is an illness that can't be fought by one person alone. One needs a network of support from people who have experienced it. Without such feedback and support, I will slip and I won't have any tools to deal with my cravings. I find I must be certain to eat the right foods in addition to being careful not to eat the wrong ones. I find this central in reducing my food cravings.

Even though I have slipped, I still see great results from Sunrider. First of all, there is no comparison to the way I feel with the herbs than the way I do without them. If I feel gassy and acidic, a cup of Fortune Delight and Calli tea will break it up in a matter of hours. Also, I am experiencing significant attitudinal changes. I have a new willingness to dive into life and tackle it that was not there before.

I am therefore achieving more and feel a new success from life for which I am deeply grateful.

I wish to comment on the psychological component of illness. In my own case, I had always been rather withdrawn from people and from life. There came a point, just before my digestive disorders began, when I abandoned my music — the most important and nourishing part of my life. This was when my health problems increased dramatically and the real drama of my health odyssey began. I also know that as I do what is important to me and even think thoughts that are healthy for me, abdominal tension lessens and the feeling is generally lighter.

For all those who are either ill or well, I say live your life with joy, imbibe its riches and do all that is most fulfilling to you. This will make you well. In particular, I wish those who are ill with Candida a speedy recovery and a joyful and fulfilling life.

A MODEST TESTIMONY

"Margaret" is a musician, mother, and dancer who works as a waitress and masseuse/bodyworker. She has been actively interested in holistic health issues since 1970.

My story is not one of spectacular recovery from a host of yeast-based illnesses. In fact, my story is probably more like the average person who suffers from several minor ailments, still carries on a normal existence, and ascribes occasional low grade upsets to the vagaries of approaching middle age.

I have always enjoyed very good health. I'm one of those people who should receive attendance awards for not missing a sick day from work in five years. The health problems I have been aware of always seemed like "emotional" health problems rather than physical: chronic difficulty controlling my weight (and the attendant anxiety over it); periodic bouts of depression; and an itchy scalp condition that was clearly stress-related. These were problems of many years' standing, but I was convinced that they would go away if I could only get my emotional life in order. Interestingly, they are also very typical "women's" health problems.

The past year has been a critical juncture in my life. My husband and I spent six months in marriage counseling and decided, after eleven years, to separate. Knowing that the period of moving to an apartment would be hectic, I started using some herb food supplements, on the advice of a friend who told me they would provide extra stamina. I say "supplements" because that's what I was expecting, even though the box said "whole food." In fact, I used the herbs almost exclusively as my entire diet in the beginning because it was several weeks before I had my kitchen supplies unpacked.

Suddenly I had the most dramatic weight loss I have ever experienced. It wasn't surprising that I was losing weight, since I was hardly eating and was under tremendous pressure, in addition to getting almost no sleep. What was surprising is that I felt wonderful. I rationalized: "You needed to get out of that relationship; you feel liberated and exhilarated..." Actually, that was the truth, in part. But I kept waiting for the letdown, the fatigue, the wasted, rundown burnout. It seemed as though I'd been running on empty and living on spent reserves. I was bound to come crashing from my high place and terrific feeling of well-being and then be exhausted and sick. But it simply didn't happen. In four weeks I had lost 16 pounds — more than I'd lost before on a total fast. As an avid exercise and dance devotee, I could see that I wasn't losing any muscle mass at all, only fat. Nor was I fatigued in class. The Taiwan flu epidemic ran its course at my workplace. I was the only person there not stricken by it.

Since I was consuming a large quantity of the Sunrider herb food formulas, which significantly contributed to my good health over the last few months, I decided to become my own wholesale distributor. At a distributor's meeting I met a nutritional consultant. He advised me that although I had a sturdy constitution, I was afflicted with yeast overgrowth (also called Candida). I had once noticed a book title in a health food store with YEAST in boldface print, to which I remember responding with instant aversion. It sounded so distasteful! What fastidious person who doesn't have vaginitis, thrush, or athletes foot wants to think her major bodily systems are riddled with an organism that is neither animal nor vegetable but sounds vaguely unclean and moldy? I'd had the same reaction when I saw the book at a friend's house. I didn't want to hear about it. I didn't want to discuss it. This nuitritionist was telling me something I wasn't psychologically prepared for.

A few days later I saw a questionnaire in a health food store entitled, "What is your Y-Score?" It was a checklist to determine whether you may have, probably have, or undoubtedly have a yeast problem. Although some expensive lab tests are available, the most reliable indicator of positive diagnosis at the present time is the questionnaire (along with observation by a practitioner experienced in dealing with this problem). Out of a possible score of 300, a score of 170 or more indicates the definite existence of a yeast problem. My score was 254.

I let all of this disturbing information sink in. Then I was ready to attack the problem in earnest. My health advisor was involved with Sunrider because he was convinced it was more effective in combatting yeast than anything else. My program consisted of continuing with the Chinese food herbs, but also altering my diet to an anti-yeast regime. My counselor followed a holistic approach which focused on rebuilding the immune system, detoxifying the digestive tract and recolonizing the intestinal floras (acidophilus, bifidus and streptococcus faecium) which had been ravaged and depleted by yeast overgrowth.

All of this was accomplished through the use of the specific formulas for each organ system that was in need of regeneration. As I researched the yeast issue, I began to identify more and more symptoms which had previously seemed unrelated. The amazing discovery for me was that many of these symptoms were beginning to clear up as I read about them and made the connection with my own health profile.

For several years my menstrual cycle had been three weeks long instead of four. Doctors told me it was probably stress-related, but hadn't otherwise seemed concerned. After a month on the new program, my periods started coming every four weeks. And while I hadn't thought of myself particularly as a victim of premenstrual syndrome, the familiar bloating and pelvic/vascular congestion normally accompanying each cycle were suddenly gone. I am still finding myself caught by surprise each month at the onset of menstruation — so absent are any of the usual PMS symptoms.

Another physiologic quirk was my tendency to feel cold or become chilled. I had always attributed this to poor circulation — an odd conclusion for someone whose work is physically demanding and who takes exercise classes every day. I believe now that my basal metabolism was low. I no longer have trouble staying warm, and my appetite has increased while I've continued losing weight. I'm definitely feeling like a more vital physical being.

One of the most dramatic changes has been in the character, tone and quality of my skin. I hadn't realized how pasty my complexion had become. I was aware of puffiness, but that I attributed to being overweight. I was also aware that if I drank so much as one beer, my face would be irritated and inflamed the following day and I would be filled with cravings for sweets. I had begun to wonder if I was allergic to alcohol. I seemed to be picking up slight allergies to almost anything, because my face would take on that same inflamed, blotchy appearance when I was exposed to barbeque smoke, tetracycline, various plants and certain food additives. During the period of apartment hunting and custody arrangements, I had so many hives and welts that I began to imagine scabies or shingles. My doctor, as always, said "Stress." Well, it was certainly a stressful time for me. But what kind of explanation was that? I saw an allergist who told me, "It's not allergies, it's immunologic response to a variety of environmental factors." The distinction had no practical relevance. There wasn't anything I could do either way. I know now that yeast disturbances cause all sorts of skin outbreaks, stress responses, and allergic reactions, which can actually be one and the same thing. This has to do with the yeast's interference in the immune system, which may manifest as the body succumbing to illness or as allergic/immunologic response, in which the immune system treats the body itself as the enemy. Premenstrual syndrome, another effect of yeast overgrowth, is also tied into this immune system pattern of disturbance; in this case it is the ovaries which are treated by the immune system as invading pathogens. Hence the intense pain some women feel during ovulation as well as menstruation.

Most of my allergic reactions, including hives, itching, asthma, swelling and acne-type rashes, have disappeared. Gone also is the panicked feeling of an anxiety attack, the edema in my knees, poor bladder control, hearing loss, deadly lethargy, toe-nail ridges, and the periodic emotional state of going almost over the edge. You may think that this range of symptoms and their seemingly miraculous cure is simply a response to feeling better in general; but perhaps the Sunrider products and the anti-yeast regime provided the basis for a change which acted merely as a catalyst for many other changes. However, since these changes began taking place, I have seen or read of every one of the symptoms mentioned above as yeast related.

My scalp no longer drives me wild at all hours of the day and night with itching. My face has finally lost that bloated, pasty look. The skin is firmed, velvety and glowing. I was amazed at how noticeable this was to other people. My weight loss (25 pounds in two months) has been the attention getting factor, but almost as frequently people comment on how my skin has improved. And sure enough, if I go off my diet, I can see it in my face the next day. Unfortunately, what this means is that it's a long uphill struggle. But the fortunate part is that I feel so much better physically and emotionally, the incentive is there to stay with the regime.

Because the die-off of large numbers of yeast releases extremely toxic byproducts, some phases in the healing process can be difficult. One of the most beneficial aspects of my counselor's approach is its holistic character. The work of killing the yeast has been balanced with the regenerative nutritional support of the Chinese whole food herb formulations. This has enabled me to feel consistently better as I recover, instead of being subjected to the overwhelming lethargy and depression so common for patients receiving standard Nystatin therapy without any other supportive therapy.

GLAD TO BE LIVING AGAIN

"Sharon" is a middle aged woman who lived an average American life with her husband and children in a suburb of Chicago. Over the years, she developed common allergies to pollen and mold and pursued the conventional medical course of treatment. One night she awoke with severe hives and difficulty breathing. She was rushed to the hospital and given several medications. The side effects from these medicines broke down her immune system function and sent her body on a two year roller coaster ride which nearly destroyed her sanity and life.

In order for me to live, my body must interreact with chemicals, food, air, water, light, temperature, drugs, dusts and pollens. In other words: everything that the 20th Century is all about.

A normal body is able to cope with this total load without becoming severely ill. For some, however, the load becomes too large and their bodies can no longer adapt. These people have weakened immune systems. I believe many people have first-stage mucocutaneous Candidiasis. Most people go through life never advancing beyond this stage. It is so prevalent it is pretty "normal" in modern day life. The first stage is conventional allergies to dust, foods, inhalants, chemicals, etc.

The second level involves more generalized reactions: pain, headaches, fatigue, rashes, joint pains and muscle aches. Third and fourth levels involve mental and behavioral responses. Sometimes there are cognitive problems: when you can't think clearly, can't think of the right words to say something, mental confusion, and so on.

This stage often takes on more severe forms of depression and for some people, schizophrenia, psychosis, irrational thoughts, panic, anxiety attacks, muscle twitching, violence, aggression, or epileptic seizures. The fourth stage experience is a virtual shutdown of various organ systems in the body. I sympathize with the people suffering from so-called mental illnesses who are institutionalized. Their problem may be Candida-related and caused by food or chemical sensitivities.

My problems aren't really allergies (a misleading word, at best), but I am sensitive to various drugs, chemicals, pollutants and food additives. These have caused my immune system to break down — it simply can't handle them. Its functions have become partially paralyzed.

How did this happen? A yeast has overpopulated and overtaken my body. It is present in my tissues, organs and bloodstream. It is called Candida. The yeast begins to populate our bodies shortly after we are born. As we grow and age, its numbers multiply and spread. If the immune system is strong, the yeast remains at this stage and is kept under control. The Candida grows when we take antibiotics, birth control pills or steroids (cortisone and Prednisone). It also grows when we eat lots of sweets (refined sugars) and high-carbohydrate foods. Candida can invade the whole intestinal tract from the mouth to the rectum and leads to a variety of symptoms.

I had all of these factors provoking the onslaught of systemic Candidiasis. I have frequently been on immune-suppressing drugs, having been treated with many doses of antibiotics for chronic recurring cystitis and for sinus problems. And I have always been a big sweet eater. I loved soda pop and cheese and would frequently indulge in high carbohydrate snacks like potato chips and french fries. All that I am sure contributed to my Candida overgrowth.

I developed the orthodox allergies: trees, grass, molds, hayfever, etc., a telltale sign that my immune system was slowly losing control over the yeast. It is quite possible that I could have gone on for the rest of my life struggling along with these more conventional allergies, without the Candida hitting as severely as it did.

However, I awoke one night in September 1985 with severe urticaria (hives) all over my body and difficulty breathing. I was rushed to the hospital where I was promptly treated with adrenaline, antihistimines and Mederol. Although I had suffered with allergies for many years, this was the first time I had reacted with urticaria. Within a few days I developed a severe vaginal yeast infection. This time, instead of using an anti-fungal drug, I was given Sultrin creme, a triple antibiotic.

Now my immune system had to contend with the existing Candida and a yeast infection as well. In the next few weeks I developed joint pains, migraines with vomiting, bruising, mental confusion and extreme fatigue. My health continued to deteriorate.

In mid-November I was again hospitalized for my urticaria. Numerous specialists were called in to diagnose my condition and I was tested for every disorder imaginable. (Chronic Candidiasis tends to mimic many disorders.) After five days in the hospital, I was transferred to a more prominent research hospital. Here I was told, "You will have the best doctors in the country to help you with your urticaria."

Within a few hours I was diagnosed as having serum sickness, a condition one can develop from a drug reaction. It took six to eight weeks for this problem to work itself out of my system. I now realize I should have thought twice about this diagnosis as most of the specialists who saw me in the next few weeks had never heard of this condition. I couldn't even find it in the medical books at the library. I was put on a therapeutic dose of the steroid, Prednisone. This was to be tapered off over the six to eight week period.

My hives were disappearing so I was discharged to recuperate at home. Within a few days I developed a low-grade fever and tachycardia. There was a constant suffocating feeling in my chest and I ached all over. I developed a numbness in my extremities, mostly my arms. At one point the paramedics rushed me to the hospital because I passed out.

I was complaining of shortness of breath, so a blood-gas test was ordered. Within minutes the test came back indicating poor results: too little oxygen and far too much carbon dioxide in my blood. I was immediately rushed to X-ray for a possible pulmonary embolism (blood clot in the lungs). It came up negative. I was in a small hospital and they told me they could not take care of my respiratory failure, so I was transferred to a larger hospital. A repeat blood gas test was ordered and it was within normal limits. I was treated for hyperventilation and sent home.

In the ensuing days it was not uncommon to see me in the emergency room of a local hospital more than once a day. At this time I developed a severe bladder infection and was put on antibiotics with Prednisone — for my so-called serum sickness. I'm wondering how I ever stayed alive, because I was so depressed and suicidal. I was feeling so sick at times that I had a death wish.

The combination of steroids and antibiotics did a very destructive job on my already weakened immune system. I soon developed severe food sensitivities. The internist diagnosed it as lactose intolerance, because every time I ate I developed tachycardia and a flushed face and became violently ill. I went through hell getting tested by all of those doctors. And most of the tests came back normal — but yet I was so ill.

Within the next ten weeks the Prednisone was

tapered. My body had become Prednisone dependent. The doctor tried to cut down the dosage three times, and each time the hives would rapidly return. And so in January I was once again admitted to the hospital for urticaria. This time it was decided that I didn't have "serum sickness" after all, because it was lasting too long.

Numerous specialists were called to help find an answer to my chronic urticaria. I saw an allergist, a dermatologist, infectious disease specialist, cardiologist, internist, gastroenterologist. And finally — an endocrinologist, for I had begun to lactate. I had not had a baby in 10 years! Candida can disrupt the fragile female hormone system.

It took days to get my hives under control this time because the adrenaline and antihistamines didn't phase them. What do you suppose they did? Increase the Prednisone, of course.

People often wonder why I don't continue my education in Nursing, since I have finished three years of a four-year degree. My experience has made me think doctors aren't God. And I can't be a hypocrite and put massive amounts of drugs into people so sick like me and expect that to be the answer. I learned from first-hand experience that drugs made me worse! By increasing the Prednisone, the insult was too great for my totally weakened immune system and the Candida totally ravaged by body.

At this stage of my illness, not one of those specialists recognized Candida Albicans as an underlying cause of my problems. I began to feel as though my throat was closing shut. I had a constant suffocating feeling in my chest and started to salivate heavily.

In this weakened state, the doctors felt they had done all they could to help me and I was discharged once again. They told me point blank: "Sometimes medical science does not have all the answers. If you are having a hard time dealing with your chronic urticaria, then perhaps you should seek psychiatric help to learn to deal with it."

Hell, I didn't need a psychiatrist to tell me I was sick, I needed medical help! At the time of discharge I had an elevated white blood count of 16,800. In orthodox medicine normal is around 9,000-10,000. My liver function tests also showed abnormal. I truly believe, when I look back, that within a short period of time I probably would have slipped into a drug-induced coma from all these drugs they were giving me. And then at this stage of the game they were giving up on me!

The very fact that an infectious agent like Candida is able to persist in my tissues indicated the inadequacy of immune response. This condition is referred to as immunologic tolerance, immunologic paralysis, and immunologic unresponsiveness. It means the Candida is now being tolerated in the tissues and implies the development of a paralysis — or unresponsiveness — of the immune system to it. This seems to happen when Candida invades the tissues to such an extent that the immune response virtually disappears. In other words, the poison (antigen) released by the Candida so overwhelms the immune system, it cannot produce an antibody to fight the Candida.

This disease — unlike any other in the traditional medical world — is highly individual because each of our bodies is unique. This is one reason the medical community has such a hard time accepting Candidiasis as a valid medical disorder. The symptoms mimic many other disorders. Too often, in Candida cases like mine, the cause of illness is never found. I believe the main cause is the over-use of prescribed antibiotics and other drugs. Orthodox medical doctors are accustomed to prescribing the same treatment or drugs for everyone with similar symptoms. Each of our bodies is different and each may exhibit different symptoms, which means we should be treated on an individual basis. Doctors are good when it comes to trauma (broken bones, for example), but when it comes to the immune system they don't seem to know how to treat or diagnose it.

You can't imagine the torment and frustration of being so sick and not finding any answers. I felt totally helpless. Finally, through sheer desperation, within 24 hours of my hospital discharge I found the help of a prominent clinical ecologist, Theron Randolph, M.D. What did I have to lose? I knew I was dying. Regular doctors weren't helping me and Dr. Randolph had a different approach to treating allergies.

Clinical ecologists are trained to look at the whole picture in an illness. Dr. Randolph didn't need to do laboratory tests. He knew by looking at my physical condition and my case history that I was in pretty bad shape. He said that I was too sick to be treated on an out-patient basis, and there was no longer a facility in the Chicago area that could treat anyone as sick as I was. So, I was flown to the Northeast Community Hospital in Bedford, Texas, that very same evening.

Dr. Randolph felt I had become a victim of environmental illness and was suffering from systemic Candidiasis. I needed a chemical-free sanctuary. The ECU (Environmental Care Unit) at the hospital has filtered air, porcelain floors, chlorine-free water and many non-pollutant items. All the nurses there have experienced at least a mild form of environmental illness.

I was what you would call "masked." In other words, my body had been exposed to so many chemicals, pollutants, drugs and foods that the reactions weren't clear. To become unmasked, I fasted on distilled or spring water in the relatively chemical free environment of the ECU for five days to clear my digestive tract of all foods. During this time my symptoms became worse for a few days before getting better.

When I became symptom-free (sometimes called baseline symptoms), I was tested for several chemicals, inhalants like grasses, weeds, pollen, molds and foods. I was started on a therapeutic dose of Nystatin and acidophilus supplements. I was off all Prednisone within five days without the return of hives!

I gradually regained my health and was released

within a couple of weeks to go back home. For the first several weeks I had to wear a charcoal mask because of chemical sensitivities to the formaldehyde building supplies which were in my new home. The detergent aisle in the grocery store made me feel dizzy and weak. I was instructed to live in the most chemical-free environment possible, to eat all natural food -- free of chemicals and preservatives — and drink only purified water. Perhaps in about three years, I was told, I could restore my immune system.

My body had reached a point where it had rejected all forms of antibiotics and many other drugs. This caused me a great deal of anxiety. What if I got sick or needed surgery? Fortunately, after being in the ECU, I realized that there are many natural remedies for the body that are just as effective as medicines and don't cause harm. This is when I realized Sunrider products could help me. I had been introduced to them in March 1986.

I started drinking the Calli beverage. Within a few days I developed intense heat and itching all over my body; then a blister-like rash on my arms, breasts and face. I also developed extreme edema in my left arm. I had a large sore which drained profusely out of the front of my neck. These blistered areas drained for about three weeks. (I believe if my immune response was stronger this would not have lasted so long.) I evacuated large amounts from my colon as well. I immediately stopped using the Calli.

Dr. Randolph said this was the worst case of inflammation he had seen in 30 years. He would not put me on antibiotics as that is what got me into this situation in the first place. He ordered massive amounts of buffered vitamin C and rest. I also took large amounts of Sunrider Alpha 20C and Conco, along with Goldenseal, during the next four weeks. Gradually I began to feel stronger and better. So I decided to introduce the Sunpack into my health program. I was able to tolerate my environment a lot better after only a few short weeks.

In mid-June I became concerned. Some of the old symptoms were recurring and were very persistent, especially the numbness and tingling in my arms and head. My mood swings were uncontrollable and erratic. I had PMS and was a true maniac at that time of the month. My doctor advised me to have the mercury amalgam fillings removed from my mouth. He told me that mercury inhibits enzyme reaction, disturbs absorption and utilization of nutrients and produces electric currents in the mouth. These currents interfere with the nervous system, altering the ration of T-cell lymphocytes, which are vital to immune defense. Worse still for us Candidiasis sufferers is that when the mercury is leached out of the amalgams, it is converted to methyl mercury. This substance disturbs the intestinal flora so that Candida Albicans is encouraged to grow. Research has shown that when Candidiasis patients fail to get well on anti-fungal therapy alone, they sometimes recover after the source of mercury intoxication is eliminated.

After having all of my mercury amalgams removed, I went through a period of withdrawal for about six to eight weeks. During this time I continued using Sunrider products, especially drinking the Calli (which is supposed to remove metal from the body). My progress has been a steady upward climb up to the present. This is the first time in 14 years I do not have to rely on allergy shots, especially to survive the ragweed season last Fall.

My skin has cleared up. My hair seems to feel shiny and full. It was falling out last Spring. My body has begun to restore a lot of the damage to the immune response which had been heavily suppressed due to all the drugs. I also develop healing reactions every so often. My reactions are evasive when I try to put a title on them. Sometimes they are emotional reactions, sometimes putrid body odors, pains, rashes. Around the time of my monthly cycle I get a muscle ache on my left arm in the same spot — month after month. Along with the muscle aches are tingling sensations, which I think is nerve damage restoring itself.

These reactions come and go and last anywhere from a few hours to a few days. They come on all of a sudden. But each time they pass I feel better. It is truly amazing what your body can do to regenerate itself, providing you take care of it — by giving it the proper nutrients.

I now realize the force of my health problems and just where they were coming from. I know how to read my body fairly well. When I am tired I lie down.

You and only you can be responsible for your own health. Since being on Sunrider and restoring my own health, I have had an indescribable feeling of well-being, vital energy, endurance and lots of strength in both my mind and body. It's almost like a revelation to me. It gives me great joy and happiness just to be alive. I will never ever take my health for granted, because if you lose it, you don't have anything.

I now realize that you can't buy health in a drugstore bottle. It is possible to heal the body, but you must cleanse the body to compensate for the wrong you have done to it during your life.

I don't think of Sunrider as a "cure" or "remedy." It is regeneration, a thorough house-cleaning. It is not the disease but the body that heals. While mending the body that has been damaged, we must allow sufficient time. After all, we didn't get sick overnight. Natural products help to heal slowly but surely. If we do our part, the body will do the rest. It seemed as though it took a long time to get well. But the constant changes and mending of my body have been so dramatic with the Sunrider products that it is an intriguing experience to me every day. The time has passed quickly and more importantly, I am so happy to be alive!

I'm very thankful that my life has been spared.

12

THE HEALING PROCESS

The healing process for any internal yeast infection goes through three phases. Each phase has specific characteristics and requires different care and remedies to facilitate the body's processes. These are: 1) cleansing, detoxification and die-off; 2) stabilization and; 3) recovery and regeneration.

The first phase can be traumatic for moderate to severely affected persons. An emotional and spiritual crisis may arise as the individual begins to look at and resolve the deeper issues and disorders that initiated the illness. It is usually a period of great fatigue or depression, sometimes with suicidal feelings. Anxiety attacks, crying, irrational anger, rage and irritability are common. All of this can be overwhelming, and one needs a good support system to get beyond this phase.

Sometimes cravings for sweets become as obsessive as the cravings of an alcoholic or drug addict for their substance. This is caused by what might be called a "rebellion of the yeast." Somehow the yeast's requirements for survival infiltrate the nervous system, and its needs become those of the individual — or so it seems.

To alleviate the physical trauma of the cleansing, one should get plenty of rest, drink lots of fluids and teas, take daily baths and occasional saunas and steam baths, and get regular, gentle exercise. A good massage can be very healing at this time as well.

Castor oil packs[1] over the affected area can help if there are cysts, growths, liver congestion, deposits breaking down, or noticeable toxicity.

At this time begin using the two Chinese health strengthening formulas (up to 16 capsules each daily), chelated vitamin C, cranberry juice, germanium and B vitamins to tolerance, plus large doses of beneficial intestinal floras. This regimen will greatly facilitate the cleansing and provide considerable relief as well. Some people experience several days of lower back pain around the kidneys as they are overworked in removing the toxins from the bloodstream.

Other physical reactions to cleansing which may be experienced during phase one are flu-like symptoms, headaches, muscular aches, boils, diarrhea, constipation, gas, and extreme fatigue.

Finally, a spiritual understanding is fundamental to complete the healing process. It involves being able to forgive all who may have abused you, intentionally or unintentionally, and a resolve to live ethically and to seek an understanding of the order of creation. Taking a clear stand on protecting the integrity and sanctity of one's space and being, and the ability to extend this to others, is essential to the healing and recovery of one's health and heart.

The stabilization phase comes when most of the heavy cleansing and trauma have passed. Recovery is just beginning. It is a time of waiting, of becoming strong and consistent in the attitudes, habits and health practices that are necessary to become and remain well. This can be a very flat time when nothing seems to be happening either way. Duration of this phase varies greatly according to the severity of the case, the individual's basic constitution, and how consistently the first phase was pursued. It can take a few days or several months.

Phase three — recovery and regeneration — is marked by increased appetite, renewed enthusiasm and enjoyment of life, more stamina, increased energy and a healing of the specific disorders. This is the most difficult phase in maintaining the diet and healthful living. The tendency may be to become careless, eating too broadly, gratifying the impulses that were deprived in the past by illness, working too hard, abusing others. It may be the first time in years that the patient has felt "normal," and zest for life takes over. It is easy to fall into bad habits.

Maintaining the diet and conscious living must continue. Otherwise, within a few weeks or months relapse occurs and the patient once again finds himself flat on his back — with another chance to learn the lesson he keeps trying to escape.

Eventually, we learn to seize the freedom that is our birthright, to live in the respect that is our deepest pleasure. And we will laugh and sing and dance and work to spread this happiness throughout the entire universe.

---

1. Instructions for preparing castor oil packs: First, cut a piece of cotton or cotton flannel large enough to cover the affected area, then, in a saucepan, heat a small amount of castor oil to a nice warm temperature and saturate the cloth. Remove, and open it up on a double piece of saran wrap, and place the cloth on abdomen or appropriate area of body wishing to treat. Place heating pad over castor oil pack for approximately one half hour while lying down. For further instructions see *The Edgar Cayce Handbook for Health.*

13

YEAST DISORDER PROTOCOL

CHINESE FOOD HERB FORMULAS

1. Cleansing and energizing tea Formulas 1 & 2: one to seven cups daily. It is very important to begin with a dilute amount: one tea bag or one teaspoon of powder to a quart of water.
2. Nourishing formula: three level tablespoons two to four times daily in water, yogurt, or vegetable broth.
3. Health Strengthening formula #1: six to sixteen capsules daily.
4. Health Strengthening formula #2: six to sixteen capsules daily.
5. Goldenseal extract: four to eight capsules daily.
6. Glandular formula: six to twelve capsules daily, depending on weakness of glands.
7. Glandular formula for women: one to three tablets daily.
8. Korean White Ginseng extract: three to six cap-

sules daily for men who need extra glandular support.

9. Digestive formula: four to twelve capsules daily with meals.
10. Fat metabolism regulation formula (also good to strengthen, heal and cleanse the digestive tract): one to three capsules of each formula two to three times daily.

BENEFICIAL INTESTINAL FLORAS

First two weeks: Streptococcus Faecium, four to eight capsules with each meal.

Second two weeks: Acidophilus, half teaspoon three times daily in four ounces warm water.

Third two weeks: Bifidus, half teaspoon three times daily in four ounces warm water.

Repeat entire procedure until recovery is complete.

BASIC VITAMIN AND MINERAL FORMULA:

Individual needs, sensitivities, and deficiencies vary greatly. Each person's condition should be evaluated individually, and a program can then be established accordingly.

OTHER FORMULATIONS AND MODALITIES:

According to individual need and preference.

14

BELIEFS, VALUES AND EXPERIENCES IN SELF DEVELOPMENT

Everything that happens to us in our lives is part of a process by which we find and come to know our true selves, our deepest needs. Again and again we are brought face to face with a reality that brings us to really love ourselves, to respect, value and care for ourselves, our creator and our creation.

A major part of getting well involves learning from observing the process of cause and effect in life and nature. By seeing how the laws of creation operate we learn to live from an experience of love and truth, and from a growing knowledge and respect for the order and function of nature. This helps us to develop our beliefs and values to find which approach brings us the fulfillment and happiness in life that we are looking for.

A violation of the order of nature is the root cause of most illnesses. Some people may go through difficulties so they can be of service, to build their character, to strengthen their faith and to learn compassion and humility. But most other illnesses are brought on by disharmonious thoughts and actions. Every person has direct access to his higher self. He has only to ask where he is making a mistake and he will be shown an answer and solution.

A complete elucidation of how violations of the natural laws and order interfaces with the development of yeast infections and disease in general is far beyond the scope of this work. It is essential to illuminate its basis, however, because no cure will last unless there is a change of heart and a change of living.

One interesting perspective on the function of yeasts is that they are parasites, and a person who lives parasitically or allows others to exploit or abuse him opens himself to being invaded by parasites. Yeasts are also scavengers, and a part of the recycling and composting phase in the ecological chain. People who live hedonistic lives, pursuing short-term gratification instead of long lasting satisfaction, are showing a lack of respect for the productive cycle and this causes them to become compost.

Yeasts also thrive where there is decay, in people who eat decayed or adulterated foods. This creates toxic waste which is fertile ground for the growth of disease. Nature nourishes those who respect and nourish her. Those who have no regard for the quality of food they eat, and give no appreciation, respect or care for the gift of their human bodies and this creation, will experience the results.

Another cause of yeast disorders is narrowness and intolerance. People who are exclusive, narrow or intolerant toward life and others will reap some form of illness, difficulty or misfortune to help them open up and learn to share love so they will be able to enjoy more fully the beauty, trust and happiness in life.

Each person has his own set of learning experiences to further his growth so he can better create a life of true love and happiness with whatever kinds of talents and handicaps, fortunes and misfortunes he is given. We limit the amount of happiness and pleasure we have in life by looking for them in the wrong places. We lose our strength and good fortune when we have a poor sense of self worth and we lose our personal power when we abuse others or when we allow others to abuse us. We need to reclaim our bodies, our lives and our consciousness from the fears, angers, and greeds to which we have given our energy. We need to love ourselves enough so that we can create the kind of world we want to live in and not just let our lives happen to us.

It is not an easy thing to live in the world today and every person who is trying to live consciously should be given support and encouragement. Many good people are being subjected to difficulties that are not of their own creation and they should be given all the help they need to overcome these difficulties. Perhaps these ancient formulas have been offered at this time to help balance the scales of justice, to make people stronger so their work will be easier in building new lives and a free and beautiful new world. Apart from any modality or therapy, genuine lasting recovery from illness will be based upon love and a recognition of personal responsibility in the process of creating reality.

15

LIFESTYLE

These are the basic factors involved in the process of recovery from illness: learning to love, respect, and care for one's own needs, learning to love and respect other people and all of creation, learning what one's limitations are, and learning to live in balance and take responsibility for one's life. The right lifestyle comes from an internal experience of truth and by living with integrity.

People who have severe yeast infections will have glandular weakness. To heal the glands requires minimizing stress. This means removing as many sources of environmental toxins and stresses as possible.

Learning to live simply and to find and nourish the intrinsic peace and happiness that is within one's being will cure the needless pursuit of pleasures and possessions that drain our energies.

It is essential for the lasting recovery from any

illness that each person find the value and beauty of his life, to learn to love and care for himself. This discovery of the intrinsic value of human life and the critical need to preserve one's freedom, leads to a deepening respect for the laws and order of creation; one begins to learn to live in balance.

Discovering what you really enjoy in life and pursuing it can be a miraculous experience. Learning to conserve one's energy and to live within one's limitations is also an essential part of this process. It is very helpful to live in a warm, clean environment and spend time in the fresh air and sunshine. Another issue is being able to identify and avoid people, situations and environments which have a negative effect on one's well being.

Sometimes experiences, situations or substances which appear as pleasures actually harm us. A lot of what society tells us to do is actually very damaging and people need to find and trust what really works for them. In this regard, the best guide is one's own personal truth, experience, observation and intuition. We need to learn to tune in to our inner self, to eat when we are hungry, rest when we need to rest, exercise as we are able. And above all, we need regular quiet times alone, getting in touch with our true selves, developing our being and purpose in life. These are all vital needs and we must give them priority in our lives.

16

CONCLUSION

The purpose of this book has been to offer a basic understanding in how to nourish and regenerate body systems that have been compromised by a yeast overgrowth illness. These include the successful use of certain Chinese herbal formulas in combination with proper consciousness, lifestyle, diet and the reimplantation of the body's natural floras.

The book is meant to provide a basic guideline for dealing with yeast conditions. The approach presented here has been most successful when applied consistently. It is not meant to be the one and only successful modality. Yeast disorders can respond well to other healing agents and approaches, and each person is wholeheartedly encouraged to seek the method that works for him– and especially to find a physician or other practitioner who is knowledgeable in treating yeast conditions.

A final word on the herbal concentrates. They work well on their own and quite well as outlined in this book, but have also been used very successfully with other modalities not discussed in these pages.

BIBLIOGRAPHY

Yeast Function and Pathology

Campbell, Mary C, B.A. &
Stewart, Joyce L., B.S. .... *The Medical Mycology Handbook*
New York: John Wiley and Sons, 1980

Crook, William G., M.D. .... *The Yeast Connection*
Jackson, Tennessee: Professional Books, 1984

Freeman, Bob, Ph.D. .... *Burrows Textbook of Microbiology, 21st Edition,*
Philadelphia: W. B. Saunders, 1973

Rippon, John W., Ph. D. .... *Medical Mycology*
Publisher unknown

Trowbridge, John, M.D. &
Walker, Morton, D.P.M. .... *The Yeast Syndrome*
New York: Bantam Books, 1986

Truss, Orian C., M.D. .... *The Missing Diagnosis*
P.O. Box 26508, Birmingham, AL 35226, 1985

Metabolic Function

Abravanel, Elliot, M.D. .... *Dr. Abravanel's Body Type Diet*
New York: Bantam Books, 1983

Asai, K., Ph.D. .... *Miracle Cure Organic Germanium*
Tokyo: Japan Publications, 1980

Barnes, Broda, M.D. &
Galton, Lawrence .... *Hypothyroidism: The Unsuspected Illness*
New York: Harper & Row, 1976

Bliznakov, Emile, M.D. &
Hunt, Gerald .... *The Miracle Nutrient Coenzyme Q10*
New York: Bantam Books, 1987

Graham, Judy .... *Evening Primrose Oil*
New York: Thorsons, 1984

Jarvis, D.C., M.D. .... *Folk Medicine*
New York: Crest Publishers, 1965

Kelley,Wm.J., D.D.S & M.D. ..... *Dr. Kelley's Answer to Cancer*
Kansas: Wedgestone Press, 1985

Langer, Stephen, M.D. .... *Solved: The Riddle of Illness*
Connecticut: Keats Publishing

McGarvey, William, M.D. .... *The Edgar Cayce Remedies*
New York: Bantam Books, 1985

Murray, Michael, ND. .... *Phyto-Pharmacia, Vol. 2*
Townsend Letter, June-July 1988

Peat, Ray Ph.D. .... *Nutrition For Women*
Eugene, Oregon: Blake College, 1983

Pfeiffer, Carl, M.D. .... *Zinc and Ohter Micronutrients*
Connecticut: Keats Publishing

Warburg, Otto Heinrich .... *The Metabolism of Tumors*
New York: R.R.Smith, 193l

Schauss, Alex, M.D. .... *Crime, Diet & Delinquency*
Biosocial Publications

Tintera, J.W. .... *Hypoadrenocorticism*
New York: Hypoglycemia Foundtn., 1973

Ziff, Sam .... *Silver Dental Fillings*
New York: Aurora Press, 1984

Nutrition

Abravanel, Elliot .... ibid.

Bachechi, Oreste .... *Kiva Light*
APW Publishers, Box 64l,
Solana Beach, California 92075, 1986

Bieler, Henry, M.D. .... *Food Is Your Best Medicine*
New York: Vintage Books, 1973

Bland, Jeffrey, Ph.D. .... *Intestinal Toxicity and Inner Cleansing*
Connecticut: Keats Publishing, 1987

Various .... *Newsletters*
Candida Research Foundation, Box 2719, Castro Valley, CA 94546

Fox, Arnold, M.D. .... *DLPA*
New York: Pocket Books 1985

Hoffman, Wendell .... *The Golden Age of Healing*
Ogden, Utah: Wendell Hoffman Publications, 1985

Howell, Edward, M.D. .... *Food Enzymes For Health and Longevity*
Connecticut: Omnagod Press, 1980

Hsu, Hong-Yen, O.M.D. .... *Oriental Materia Medica*
Long Beach, CA:Oriental Healing Arts, 1986

Jarvis, D.C., M.D. .... ibid.

Jensen, Bernard, D.C., N.D. ...... *The Science and Practice of Iridology*
Escondido, California: Bernard Jensen, 1974

Keyes, John .... *Chinese Herbs*
Rutland, Vermont: Charles Tuttle Co., 1976

Lesser, Michael, M.D. .... *Nutrition and Vitamin Therapy*
New York: Bantam Books, 1981

Lu, Henry C. .... *Chinese System of Food Cures*
New York: Sterling Press, 1986

Prescot, Robert .... *Chinese Keys to Regeneration*
Draper, Utah: JBE, Inc., 1984

Reid, Daniel .... *Chinese Herbal Medicine*
Boston, Mass: Shambala Press, 1987

Reilly, Dr. Harold .... *The Edgar Cayce Handbook For Health*
New York: Jove, 1982

Tierra, Michael, N.D. .... *The Way of Herbs*
New York: Pocket Books 1983

Tsung, Pi-Kwang, Ph.D. .... *Immunology and Chinese Herbal Medicine*
Long Beach, California: Oriental Healing Arts, 1986

Wolf, Max, M.D. &
Ransberger, Karl, Ph.D. .... *Enzyme Therapy*
Los Angeles, California: Regent House, 1977

Environmental Factors

Bieler, Henry, M.D. .... ibid.

Dadd, Debra Lynn .... *The Nontoxic Home*
Los Angeles, California: Jeremy Tarcher, Inc., 1986

Trowbridge, John, M.D. &
Walker, Morton, D.P.M. .... ibid.

Spiritual, Mental and Emotional Factors

Buscaglia, Leo, Ph.D. .... *Living, Loving and Learning*
New York: Ballantine Books, 1982

Cayce, Edgar .... *Edgar Cayce's Story of Attitudes and Emotions*
New York: Berkeley Books, 1984

Hay, Louise .... *You Can Heal Your Life*
Berkeley, California: Hay House, 1986

Illich, Ivan .... *Medical Nemesis*
New York: Bantam Books, 1977

Locke, Steven .... *Mind and Immunity*
Praeger, 1983

Norwood, Robin, M.F.C.C. .... *Women Who Love Too Much*
Los Angeles, CA: JeremyTarcher, Inc., 1985

Schneider, Meir .... *Self Healing - My Life, A Vision*
New York: Routledge & Kegan Paul, 1987

Siegel, Bernie, M.D. .... *Love, Medicine and Miracles*
New York: Harper & Row, 1986

About the Author

John Finnegan, nutritional and environmental consultant, has spent twenty years studying and working in the holistic health field. With a college background in the life sciences, he went on to study and work with many of this century's leading medical pioneers, and later received a degree as a Nutritional Consultant. He was a friend and student of Broda Barnes, M.D., author of *Hypothyroidism: The Unsuspected Illness,* an acknowledged authority in thyroid and endocrine glandular function. He also studied and worked with Dr. John Christopher, Dr. Michael Barnett, and in several holistic medical centers.

John Finnegan is the author of four books, including *Recovery From Addiction, Oxygenation: Vital Key To Life and Regeneration of Health.* He lectures and conducts seminars, and gave presentations at both the 1987 and 1988 San Francisco Whole Life Expos

Disclaimer

This book has been written and published solely for educational purposes. It should not be used as a substitute for a physician's advice.

If you need medical help you should seek out a physician or practitioner knowledgeable in this field and work under his direction.

The author provides this information with the understanding that you act on it at your own risk and with full knowledge that you should consult with health professionals for any help you need.

More Books from Elysian Arts:

Recovery from Addiction

A Comprehensive Understanding With Nutritional Therapies For Recovering Addicts and Co-Dependents

John Finnegan - Author

The purpose of this 190-page book is to provide a comprehensive overview of the factors involved in recovery from addictions, give an insight into the resources available for help, and provide the most complete, up-to-date information on the metabolic basis of addictions and co-dependency. This book also covers the most effective nutritional and medical therapies, (some of which are little known), which can help correct the biochemical disorders that lie at the basis of addictions.

Natural Foods & Good Cooking

Kathy Cituk - Author

Aimed at utilizing the maximum nutritional and healing properties of natural foods, this book meets a real need for information about food and cooking that is both simple and based on common sense as well as scientific fact. Besides a wealth of easy to prepare, delicious recipes, it features special sections on using oils and butter, salt, herbs and spices, and other foods that will provide you with practical guidance on how to eat well.

Yeast, Parasites, and Viruses

An Understanding With Nutritional Therapies

John Finnegan - Author

This 160-page book is filled with fascinating case histories and gives a wealth of tried and successful nutritional therapies for helping the body to recover from Epstein-Barr and other viral conditions, colds and flus, parasites, Candida, and many of the maladies which plague man today.

If you would like to purchase any of the above books,
Please fill out the order blank, and send it along with your check to
Elysian Arts 20 Sunnyside Avenue, Suite A 161, Mill Valley, CA 94941

Natural Foods, Good Cooking			Recovery from Addiction or Yeast, Parasites, and Viruses		
Quantity	Cost each	Shipping*	Quantity	Cost each	Shipping*
1 to 9	$7.00	$5.00	1 to 9	$9.00	$5.00
10 to 49	6.50	7.00	10 to 49	8.00	7.00
50 to 99	5.50	12.00	50 to 99	7.00	12.00
100 or more	4.50	20.00	100 or more	5.50	20.00

* Please note that shipping costs outside the U.S. will be higher.
Please enclose double the amount indicated. Checks must be payable in U.S. funds.

Please send ________ copies of ____________________ @ $ ________

________ copies of ____________________ @ $ ________

________ copies of ____________________ @ $ ________

Book Total = ________

Name: ____________________

6% Sales Tax = ________
(California Only)

Address: ____________________

City: ____________________

Shipping* = ________

State: ________ Zip: ________

Total = ________